Geometric Algebra and Applications to Physics

Geometric Algebra and Applications to Physics

VENZO DE SABBATA
BIDYUT KUMAR DATTA

Taylor & Francis
Taylor & Francis Group
New York London

Taylor & Francis is an imprint of the
Taylor & Francis Group, an informa business

CRC Press
Taylor & Francis Group
6000 Broken Sound Parkway NW, Suite 300
Boca Raton, FL 33487-2742

Printed in the United States of America on acid-free paper
10 9 8 7 6 5 4 3 2 1

International Standard Book Number-10: 1-58488-772-9 (Hardcover)
International Standard Book Number-13: 978-1-58488-772-0 (Hardcover)

Library of Congress Cataloging-in-Publication Data

De Sabbata, Venzo.
Geometric algebra and applications to physics / Venzo de Sabbata and Bidyut Kumar Datta.
p. cm.
Includes bibliographical references and index.
ISBN 1-58488-772-9 (alk. paper)
1. Geometry, Algebraic. 2. Mathematical physics. I. Datta, Bidyut Kumar. II. Title.

QC20.7.A37D4 2006
530.15'1635--dc22 2006050868

Visit the Taylor & Francis Web site at
http://www.taylorandfrancis.com

and the CRC Press Web site at
http://www.crcpress.com

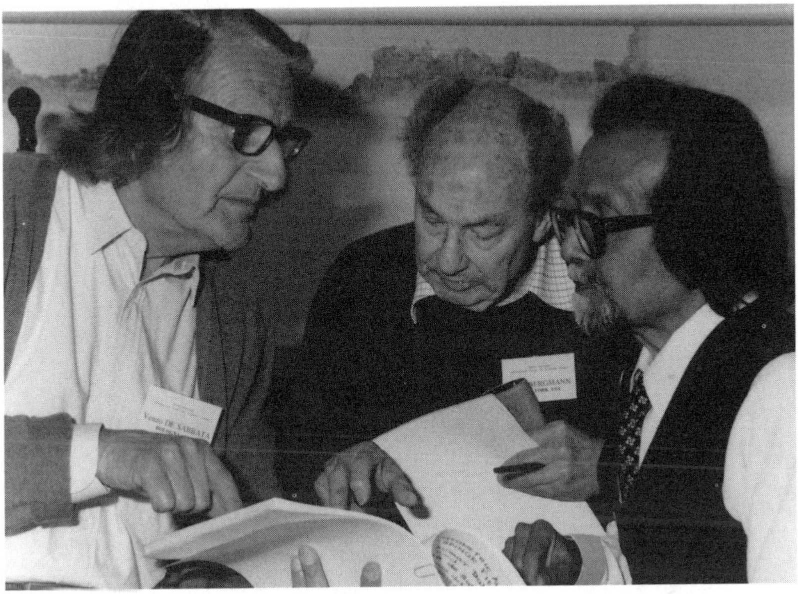

The authors with Peter Gabriel Bergmann.
From the left: Venzo, Peter, Datta.

Preface

This is a textbook on geometric algebra with applications to physics and serves also as an introduction to geometric algebra intended for research workers in physics who are interested in the study of this modern artefact. As it is extremely useful for all branches of physical science and very important for the new frontiers of physics, physicists are very much getting interested in this modern mathematical formalism.

The mathematical foundation of geometric algebra is based on Hamilton's and Grassmann's works. Clifford then unified their works by showing how Hamilton's quaternion algebra could be included in Grassmann's scheme through the introduction of a new geometric product. The resulting algebra is known as Clifford algebra (or geometric algebra) and was introduced to physics by Hestenes. It is a combination of the algebraic structure of Clifford algebra and the explicit geometric meaning of its mathematical elements at its foundation. Formally, it is Clifford algebra endowed with geometrical information of and physical interpretation to all mathematical elements of the algebra.

It is the largest possible associative algebra that integrates all algebraic systems (algebra of complex numbers, matrix algebra, quaternion algebra, etc.) into a coherent mathematical language. Its potency lies in the fact that it can be used to develop all branches of theoretical physics envisaging geometrical meaning to all operations and physical interpretation to mathematical elements. For instance, the spinor theory of rotations and rotational dynamics can be formulated in a coherent manner with the help of geometric algebra. One important fact is to develop the problem of rotations in real space-time in terms of spinors, which are even multivectors of space-time algebra. This fact is extremely important because it allows us to put tensors and spinors on the same footing: a necessary thing when we, through torsion, introduce spin in the general theory of relativity.

This later argument seems to be very important when we will try to consider a quantum theory for gravity. Moreover, the problem of rotations in real space-time allows us to explain the neutron interferometer experiments in which we know that a fermion does not return to its initial state by a rotation of 2π; in fact, it takes a rotation of 4π to restore its state of initial condition.

Geometric algebra provides the most powerful artefact for dealing with rotations and dilations. It generalizes the role of complex numbers in two dimensions, and quaternions in three dimensions, to a wider scheme for dealing with rotations in arbitrary dimensions in a simple and comprehensive manner.

The striking advantage of an entirely "real" formalism of the Dirac equation in space-time algebra (geometric algebra of "real" space-time) without using complex numbers is that the internal phase rotations and space-time rotations are considered in a single unifying frame characterizing them in an identical manner.

However, other important physical interpretations are based on geometric algebra as we will show in this book. For instance, geometric algebra (GA) and electromagnetism, GA and polarization of electromagnetic waves, GA and the Dirac equation in space-time algebra, GA and quantum gravity, and also, GA in the case of the Majorana–Weyl equations, to mention only a few.

Venzo de Sabbata
Bidyut Kumar Datta

Introduction

There are many competing views of the evolution of physics. Some hold the perspective that advances in it come through great discoveries that suddenly open vast new fields of study. Others see a very slow, continuous unfolding of knowledge, with each step along the path only painstakingly following its predecessor. Still others see great swings of the pendulum, with interest moving almost collectively from the original edifice of classical physics to the 20th century dominance of quantum mechanics, and perhaps now back again towards some intermediate ground held by nonlinear dynamics and theories of chaos. Superimposed on all of this, of course, is the overriding theme of unification, which most clearly manifests itself in the quest for a theory that fully unifies the best descriptions of all the known forces of nature.

However, there is still another kind of evolution of thought and unification of theory that has quietly yet effectively gone forward over the same scale of time, and it has been in the very mathematics itself used to describe the physical attributes of nature. Just as Newton and Leibniz introduced calculus in order to provide a centralized, rigorous framework for the development of mechanics, so have many others conceived of and applied ever-refined mathematical techniques to the needs of advancing physical science. One such development that is only now beginning to be truly appreciated is the adaptation by Clifford of Hamilton's quaternions to Grassmann's algebraic theory, which resulted in his creation of a geometric form of algebra. This powerful approach uses the concepts of bivectors and multivectors to provide a much simplified means of exploring and describing a wide range of physical phenomena.

Although several modern authors have done a great deal to introduce geometric algebra to the scientific community at large, there is still room for efforts focused on bringing it more into the mainstream of physics pedagogy. The first steps in that direction were originally taken by David Hestenes who wrote what have become classic books and papers on the subject. As the topic gets further incorporated into undergraduate and graduate curricula, the need arises for the ongoing development of textbooks for use in covering the material. Among the authors who have recognized this need and acted on it are Venzo de Sabbata of the University of Bologna and Bidyut Kumar Datta of Tripura University in India, and the publication of their book *Geometric Algebra and Its Applications to Physics* is the satisfying result.

The authors are well known for their research in general relativity. The roles of torsion and intrinsic spin in gravity have been recurring themes, especially in the work of de Sabbata, and these topics have played a central role in the interesting approaches that he, Datta, and others have taken to the quantization of gravity. He has served, since its founding, as the Director of the International School of Cosmology and Gravitation held every two years at the Ettore Majorana Centre for Scientific Culture in Erice, Sicily. It has been at these schools that many of the best general relativists, mathematical physicists, and experimentalists have explored the interplay between classical and quantum physics, with emphasis on understanding the role of intrinsic spin in relativistic theories of gravity. Datta, a mathematician, is a familiar figure at these schools, and with de Sabbata has published several of the seminal papers on the application of geometric algebra to general relativity. The *Proceedings of the Erice Schools* contain a number of their relevant papers on this subject, as well as interesting works in the area by others, including the Cambridge group consisting of Lasenby, Doran, and colleagues.

The book seeks to not only present geometric algebra as a discipline within mathematical physics in its own right but to show the student how it can be applied to a large number of fundamental problems in physics, and especially how it ties to experimental situations. The latter point may be one of the most interesting and unique features of the book, and it will provide the student with an important avenue for introducing these powerful mathematical techniques into their research studies.

The structure of *Geometric Algebra and Its Applications to Physics* is very straightforward and will lend itself nicely to the needs of the classroom. The book is divided into two principal parts: the presentation of the mathematical fundamentals, followed by a guided tour of their use in a number of everyday physical scenarios.

Part I consists of six chapters. Chapter 1 lays out the essential features of the postulates and the symbolic framework underlying them, thus providing the reader with a working knowledge of the language of the subject and the syntax for manipulation of quantities within it. Chapter 2 then provides the first look at bivectors, multivectors, and the operators used on and with them, thus giving the student a working knowledge of the main tools they will need to develop all subsequent arguments. Chapter 3 eases the reader into the use of those tools by considering their application in two dimensions, and it presents the introductory discussion of the spinor. Chapter 4 is devoted to the extension of those topics into three dimensions, whereas Chapter 5 opens the door to relativistic geometric algebra by explaining spinor and Lorentz rotations. Chapter 6 then devotes itself completely to a description of the full form of the Clifford algebra itself, which combined the work of Hamilton and Grassmann in its original formulation and was given its modern character by Hestenes.

Part II of the book then provides the crucial sections on the application of geometric algebra to everyday situations in physics, as well as providing examples of how it can be adapted to examine topics at the frontiers.

It opens with Chapter 7, which shows how Maxwell's equations can be expressed and manipulated via space-time algebra, using the Minkowski space-time and the Riemann and Riemann–Cartan manifolds. Chapter 8 then shows the student how to write the equations for electromagnetic waves within that context, and it demonstrates how geometric algebra reveals their states of polarization in natural and simple ways. There are two very helpful appendices to that chapter: one is on the role of complex numbers in geometric algebra formulations of electrodynamics and other covers the details of generating the plane-wave solutions to Maxwell's equations in this form. Chapter 9 provides the interface between geometric algebra and quantum theory. Its topics include the Dirac equation, wave functions, and fiber bundles. With the proper tools in place, the authors then go about using them to explore the fundamental aspects of intrinsic spin and charge conjugation and, their centerpiece, to interpret the phase shift of the neutron as observed during neutron interferometry experiments carried out in magnetic fields. It is during the latter discussion that the value of geometric algebra as applied to experimental findings becomes quite evident. The book ends with Chapter 10, a return to the original research interests of the authors: the application of geometric algebra to problems central to the quantization of gravity. Spin and torsion play key roles here, and the thought emerges that geometric algebra may well be what is needed to usher in a new paradigm of analysis that is capable of placing the essential mathematical features of general relativity on a common setting with those of quantum theory.

As alluded to above, it is somehow very appealing that the great quest for a unified description of the forces of nature, started by Maxwell, should have evolved towards its goal over essentially the same period of time that the mathematical unification embodied by Clifford algebra and its subsequent evolution took place. This is more than just a serendipitous coincidence, in that the past 150 years have seen a constant striving for improvements in the mathematical tools of physics, and the deepest structure of nature itself has come to be understandable only in terms of the pure mathematics of group theory and topology. We should not be surprised, then, that the very natural mathematical synthesis inherent to geometric algebra should cause it to fit so well with all branches of physics, and we can be grateful to de Sabbata and Datta for encapsulating this powerful methodology in a contemporary textbook that should prove useful to generations of students.

George T. Gillies
University of Virginia
Charlottesville, Virginia

Contents

Part I

1

The Basis for Geometric Algebra

1.1 Introduction

Geometric algebra combines the algebraic structure of Clifford algebra with the explicit geometric meaning of its mathematical elements at its foundation. So, formally, it is Clifford algebra endowed with geometrical information of and physical interpretation to all mathematical elements of the algebra. This intrusion of geometric consideration into the abstract system of Clifford algebra has enriched geometric algebra as a powerful mathematical theory.

Geometric algebra is, in fact, the largest possible associative division algebra that integrates all algebraic systems (viz., algebra of complex numbers, vector algebra, matrix algebra, quaternion algebra, etc.) into a coherent mathematical language that augments the powerful geometric intuition of the human mind with the precision of an algebraic system. Its potency lies in the fact that it develops all branches of theoretical physics, envisaging geometrical meaning to all operations and physical interpretation to mathematical elements, e.g., it integrates the ideas of axial vectors and pseudoscalars with vectors and scalars at its foundation. The spinor theory of rotations and rotational dynamics can be formulated in a coherent manner with the help of geometric algebra.

1. Geometric algebra provides the most powerful artefact for dealing with rotation and boosts. In fact, it generalizes the role of complex numbers in two dimensions, and quaternions in three dimensions, to a wider scheme for tackling rotations in arbitrary dimensions in a simple and comprehensive manner.
2. The striking advantage of an entirely "real" formulation of the Dirac equation in space-time algebra (geometric algebra of "real" space–time) without using complex numbers is that the internal phase rotations and space–time rotations are considered in a single unifying frame characterizing them in an identical manner.
3. W.K. Clifford synthesized Grassmann's algebra of extension and Hamilton's quaternion algebra by introducing a new type of product ab of two proper (non-zero) vectors, called geometric product. He constructed a powerful algebraic system, now popularly known

as Clifford algebra, in which vectors are equipped with a single associative product that is distributive with respect to addition.

Geometric algebra, developed by Hestenes [1,2,3] during the decades 1966–86, though serving as a powerful mathematical language for the development of physics, is still not widely known.

1.2 Genesis of Geometric Algebra

An account of the concept of numbers and directed numbers that had been evolving from antiquity to the 17th century, when symbolism of algebra had been developed to a degree commensurate with Greek geometry, is given with full historical background. The deficiencies in the concept of number in Descartes' time, however, were removed with the advent of calculus that gave a clear idea of the "infinitely small." A transparent idea of "infinity" and of the "continuum of real numbers" was conceived in the later part of the 19th century by Weierstrass, Cantor, and Dedekind when real numbers were defined in terms of natural numbers and their arithmetic without taking any recourse to geometric intuition of the "linear continuum." However, the evolution of the concept of number did not stop here as it would depend more on the geometric notion than on the linear continuum.

With a proper symbolic expression for direction and dimension came the broader concept of directed numbers — *multivectors* — which is a powerful mathematical language for physical theories, the *sine qua non* for future direction.

Euclid made a systematic formulation of Greek geometry (310 B.C.) from a handful of simple assumptions about the nature of physical objects. This, in fact, provided the first comprehensive theory of the physical world that led to the foundation for all subsequent advances in physics. In accordance with Plato's ideal world of mathematical concepts (360 B.C.), geometrical figures were regarded as idealization of physical bodies. The great Greek philosopher Plato (429–348 B.C.) seems to have foreseen some of the wonderful insights, such as

1. Mathematics must be studied for its own sake and perceived by the exercise of mathematical reasoning and insight; its completely accurate applicability to the objects of the physical world must not be demanded.
2. Physical theory, on the other hand, could ultimately be developed and understood only in terms of precise mathematics [4,5,6].

The mathematical concepts of Plato's ideal world were only approximately realized in terms of the observed features of the physical world we live in. The central theme of Greek geometry was the theory of congruent figures that specified a set of rules to be used for classifying bodies with a proper notion of size and shape. The idea of measurement could have been conceived after Greek geometry was created, though it was not created with the problem of measurement in mind.

In this regard we would like to be more precise and question the usual point of view according to which, in general, Hellenism appears to be a period of decline [7].

On the contrary, the birth of "modern science" goes back 2000 years, namely near the end of the 4th century B.C. The most known scientists of that time, Euclid and Archimedes (Euclid with the ability of abstraction of a thought devoted mostly to philosophical speculations, and Archimedes as the inventor of burning glass) were not the isolated precursors of a form of thought that would flourish later on only in the 17th century A.D. Instead, they were two of a large group of outstanding scientists: Erofilo of Calcedonia (around the first half of the 3rd century B.C.), founder of scientific medicine; Eratostene of Cirene (around the second half of the 3rd century B.C.), the first mathematician who gave a very precise measurement of the length of the earthly (terrestrial) meridian; Aristarco of Samo (the same epoch of the 3rd century B.C.), founder of the heliocentric system; Ipparco of Nicea (in the 2nd century B.C.), precursor of the modern dynamics and gravitation theory; Ctesibio of Alessandria (first half of the 3rd century B.C.) who developed the science of compressible fluids, as well as many others who were protagonists of a sort of scientific revolution that achieved very high levels of theoretical elaboration together with experimental practice that was not inferior to that of Galileo and Newton.

Strangely, the scientists involved in research from the Renaissance period to date seem to ignore the testimony of this extraordinary phenomenon. According to Lucio Russo [7], it appears that the Roman people destroyed the Hellenistic states after the conquest of Syracuse, the killing of Archimedes (212 B.C.) and the destruction of Corinto (146 B.C.). The indifference of Rome to scientific culture accounted for most of the original texts being lost. According to Russo [7], the birth of modern science was not an independent or a casual event; "modern" scientists gradually took possession of the branches of knowledge as they were brought to light by the discovery of the Greek, Arab, and Byzantine manuscripts.

Euclid sharply distinguished between number and magnitude, associating the former with the operation of counting and the latter with a line segment. So, for Euclid, only integers were numbers; even the notion of fractions as numbers had not yet been conceived of. He represented a whole number n by a line segment that was n times the chosen unit line segment. However, the opposite procedure of distinguishing all line segments by labeling them with numerals representing counting numbers was not possible. Obviously, this one-way correspondence of counting number with magnitude implies that the latter concept was more general than the former. The sharp distinction between counting number and magnitude, made by Euclid, was an impediment to the development of the concept of number. Even the quadratic equations whose solutions are not integers or even rational numbers were regarded to have no solutions at all. The Hindus and Arabs were able to resolve the problem of generalizing their notion of number by separating the concept of number from that of geometry. By retaining the rigid distinction between the two concepts, Euclid expressed problems of arithmetic and algebra into problems of geometry and solved them for line segments instead of for numbers. Thus, he represented the product $xx(=x^2)$ by a square with each side of magnitude x, and the product xy by a rectangle with sides of magnitude x and y. Likewise, x^3 is represented by a cube with each edge of magnitude x, and xyz by a rectangular parallelepiped with edges of magnitude x, y, and z. However, there being no corresponding representation x^n for $n > 3$ in Greek geometry, the Greek correspondence between algebra and geometry could not be extended beyond $n = 3$. This breakdown of Euclid's procedure of expressing every algebraic problem into a geometric problem impeded the development of algebraic methods. These "apparent" limitations of Greek mathematics were, however, overcome in the 17th century by René Descartes (1596–1650) who developed algebra as a symbolic system for representing geometric notions. This, in fact, led to the understanding of how subtle the far-reaching significance of Euclid's work was.

Also, here we would like to stress that the fact that limitation of Greek mathematics was only apparent and *not* real is shown by the works of Pitagora ($\sim$ 585–500 B.C.) after the development of mathematics by Talete (640–546 B.C.) and their disciples (called "Pythagoreans"). In fact, in Pythagoreans one can find a strong correspondence between mathematics (numbers) and geometry: he and the Pythagoreans have shown that the properties of numbers (for Pitagora, number means integer number) were evident through geometric disposition (observe for instance that 1, 4, 9, 16, etc., were called "squared" numbers because, as points, they can be disposed in squares). The Pythagoreans were also shocked by the discovery that some ratios (as for instance the ratio between the hypotenuse and one of the catheti or the ratio between the diagonal of a square with its side) could not be represented by integers. They were so shocked that they thought that this should not be brought to light but must stay secret! It is the first evidence of the presence of numbers with extra reason (beyond reason), and therefore called "irrational" numbers. However, what we like to stress is that the correspondence between mathematics (numbers) and geometry was already present in the old Greek science.

After the remarkable development of science and mathematics in ancient Greece, there was a long scientific incubation until an explosion of scientific knowledge in the 17th century gave birth to new science, known as Renaissance science. The long hiatus between the Greek science of antiquity and Renaissance science can plausibly be explained by its historical evolution. The evolution of science is determined by its inherent laws. The advances of the Renaissance had to wait for the development of an adequate number system that could express the results of measurement and of a proper formulation of an algebraic language to express relations among these results. During this period of scientific incubation the decimal system of Arabic numerals was invented and a comprehensive algebraic system began to take shape.

In 250 A.D., Diophantes, the last of the great Greek mathematicians, accepted fractions as numbers. In 1540, Vieta studied rules for manipulating numbers in an abstract manner by introducing the idea of using letters to represent constants as well as unknowns in algebraic equations. This, in fact, revealed the dependence of the concept of number on the nature of algebraic operations. Before Vieta's innovations, the union of algebra and geometry could not have been accomplished. This union could have been consummated only when the concept of number and the symbolism of algebra had been developed to a degree commensurate with Greek geometry. When the stage of development in two fronts — the concept of number and the symbolism of algebra — had just been achieved, René Descartes appeared on the scene.

Though from the very beginning algebra was associated with geometry, Descartes first developed it systematically in geometric language. Three steps are of fundamental importance in this development. First, he assumed that every line segment could be uniquely represented by a number that endowed the Greek notion of magnitude a symbolic form. Second, he labeled line segments by letters representing their numeral lengths. This resided in the fact that the basic arithmetic operations of addition and subtraction could be described in a completely analogous way as geometric operations on line segments. Third, in order to get rid of the apparent limitations of the Greek rule for geometric multiplication, he invented a rule for multiplying line segments, yielding a line segment in complete correspondence with the rule for multiplying numbers. By introducing a symbol such as $\sqrt{2}$ to designate a solution of the equation $x^2 = 2$, it was possible to recognize the reality of algebraic numbers. By taking recourse to the above steps, Descartes accomplished the task of uniting algebra and geometry started by the Greek mathematicians. Moreover, Descartes was able to use algebraic equations to describe geometric curves, which heralded the beginning of analytic geometry. Indeed, this was a crucial step in the development of mathematical language for modern physics. The assumption of a complete correspondence between numbers and line segments was the basis of union of algebra and geometry achieved by Descartes. Pierre de Fermat (1601–1665) independently obtained similar results. But Descartes penetrated into the heart of the problem by uniting his concept of number with the Greek notion of geometric magnitude, which opened up new vistas of scientific knowledge unequalled in the history of the Renaissance period.

In this context it is quite relevant to note what Descartes wrote to Mersenne in 1637:

> I begin the rules of my algebra with what Vieta
> wrote at the very end of his book. . . .
> Thus, I begin where he left off.

Vieta used letters to denote numbers, whereas Descartes introduced letters to denote line segments. Vieta studied rules for manipulating numbers in an abstract manner, and Descartes accepted the existence of similar rules for manipulating line segments and greatly improved symbolism and algebraic technique. Thus, it seemed that numbers might be put into one-to-one correspondence with points on a geometric line, leading to a significant step in the evolution of the concept of number.

The deficiencies in the concept of number in Descartes' time could be felt with the advent of calculus, which gave a clear idea of the "infinitely small." A transparent idea of "infinity" and the "continuum of real numbers" was conceived in the 19th century by Weierstrass, Cantor, and Dedekind when real numbers were defined in terms of natural numbers and their arithmetic without taking any recourse to geometric intuition of the continuum. This arithmeticization of real numbers, in fact, imparted a precise symbolic expression to the intuitive concept of a continuous line.

The far-reaching significance of Descartes' union of number and geometric length still resides in the fact that real numbers could be put into one-to-one correspondence with points on a geometric line. The development of algebra as a symbolic system for representing geometric notions was a great turning point of Renaissance science. But the evolution of the concept of number did not stop here, as it would depend more on the geometric notions than on the linear continuum.

Descartes' algebra could be used to classify line segments by length only. The fundamental geometric notion of direction of a line segment finds no expression in ordinary algebra. The modification of algebra to have a fuller symbolic representation of geometric notions had to wait some 200 years after Descartes, when the concept of number was generalized by Herman Grassmann to incorporate the geometric notion of direction as well as magnitude. With a proper symbolic expression for direction and dimension came the broader concept of directed numbers, now known as multivectors.

We have already mentioned that the theory of congruent figures was the central theme of Greek geometry. Descartes designated two line segments by the same positive real number, which we now call the positive scalar, if one could be obtained from the other by a translation or a rotation or by a combination of both. Conversely, every positive scalar was represented by a line segment without any restriction to its position and direction, i.e., all congruent line segments were regarded as one and the same.

In order to conceive of the idea of directed number, Herman Grassmann generalized the concept of number by incorporating the geometric notion of both direction and magnitude in his book *Algebra of Extension* in 1844. He invented a rule for relating directed line segments to numbers. In contrast to Descartes' idea, he regarded two line segments as equivalent and designated them by the same symbol, if and only if one could be obtained from the other by a translation. On the other hand, he regarded two line segments as possessing different directions and designated them by different symbols, if and only if one can be obtained from the other by a rotation or by a combination of translation and rotation. Thus, Grassmann conceived of the idea of directed line segment or directed number, called vector. A vector is graphically represented by a directed line segment and embodies the essential abstractions of magnitude and direction without any restriction to its position.

Through his revelation that the concept of number must be based on the rules for combining two numbers to get a third, Grassmann invented the rules for combining vectors, which would fully describe the geometrical properties of directed line segments. Thus, he set down algebraic rules for addition and multiplication of a vector by a scalar that must obey the commutative and associative rules such as in Descartes' algebra. The zero vector was regarded as one and the same number as the zero scalar.

In order to endow the algebraic system for vectors with a complete symbolic expression of the geometric notion of magnitude and direction, Grassmann introduced two kinds of multiplication for vectors, viz., inner and outer products. He defined the inner product of two vectors **a** and **b**, denoted by $a \cdot b$, to be a scalar obtained by dilating the perpendicular projection of a on b by the magnitude of b, or equivalently by dilating the perpendicular projection of b on a by the magnitude of a:

$$a \cdot b = |a| \cos \vartheta |b| = |b| \cos \vartheta |a| = b \cdot a, \tag{A}$$

where ϑ is the angle between a and b. The inner product can as well be defined abstractly as a rule relating scalars to vectors that has all the basic properties provided by the above definition of inner product in terms of perpendicular projection. The expression (A) abstractly calls for an independent definition of the angle ϑ between vectors **a** and **b**. The magnitude of a vector is related to the inner product by

$$a \cdot a = |a|^2 \geq 0. \tag{B}$$

In what follows, we shall show how the preceding arguments leading to the invention of scalars and vectors can be continued in a natural way, which, in turn, further extend the concept of number by the introduction of bivector or outer product of two vectors **a** and **b**, denoted by the symbol $a \wedge b$. The fundamental geometrical fact that two distinct lines intersecting at a point determine a plane, or more specifically, that two noncollinear directed line segments determine a parallelogram, was considered by Grassmann who gave it a direct algebraic expression. For this purpose he regarded a parallelogram as a kind of "geometrical product" of its sides. More specifically, he introduced a new kind of directed number of dimension two — a plane-like object — having both magnitude and orientation, such as an oriented flat surface and the rotation in a plane. It is graphically represented by an oriented parallelogram defined by two vectors **a** and **b** with the head of **a** attached to the tail of **b**, and mathematically represented by the bivector $a \wedge b$, also called the outer product of **a** and **b**. A bivector represents the essential abstractions of magnitude and planar orientation without any restriction to the shape of the plane. It is to be noted that the bivector $a \wedge b$ is different from the usual vector product $\mathbf{a} \times \mathbf{b}$, which is an axial vector in Gibbs' vector algebra.

In 1884, just 40 years after the publication of Grassmann's *Algebra of Extension*, Gibbs developed his vector algebra following the ideas of Grassmann by replacing the concept of the outer product by a new kind of product known as vector product and interpreted as an axial vector in an ad-hoc manner. This, in fact, went against the run of natural development of directed numbers started by Grassmann and completely changed the course of its development in the other direction. Grassmann's outer product reveals the fact that the Greek distinction between number and magnitude has real geometric significance. Greek magnitudes, in fact, added like scalars but multiplied like vectors, asserting the geometric notions of direction and dimension to multiplication of Greek magnitudes. This revealing feature is a reminiscence of the distinction, carefully made by Euclid, between multiplication of magnitudes and that of numbers. Thus, Herman Grassmann fully accomplished the algebraic formulation of the basic ideas of Greek geometry begun by René Descartes.

During 1966–86, David Hestenes [1–3] constructed an algebraic system known as geometric algebra, which combined the algebraic structure of Clifford algebra (1876) with the explicit geometric meaning of its mathematical elements — directed numbers of different dimensions — at its foundation. He termed these directed numbers multivectors. Thus scalars are termed as multivectors of grade 0, vectors as multivectors of grade 1, bivectors as multivectors of grade 2, trivectors as multivectors of grade 3, etc. A volume-like object having magnitude as well as a choice of handedness is graphically represented by an oriented parallelepiped with handedness defined by three vectors **a**, **b**, and **c**, and mathematically represented by trivector $a \wedge b \wedge c$. A trivector represents the essential abstraction of volume orientation with handedness and magnitude without any restriction to the shape of the volume-like object. For n-dimensional space, multivectors with grade greater than n cannot be constructed and hence they cease to exist.

In contrast to Gibbs, Hestenes retained Grassmann's concept of outer product of vectors, extended it in a natural way to get multivectors of higher grade and successfully developed geometric algebra — a powerful mathematical language for physics.

1.3 Mathematical Elements of Geometric Algebra

Geometric algebra for three-dimensional space consists of four types of mathematical elements having correspondences with geometrical or physical objects. So, the powerful geometric intuition of the human mind and the physical objects are built into its very foundation. We give qualitative ideas of these four types of elements of this algebra for three-dimensional space. Details are provided in Reference[8].

1. First, we consider physical objects having magnitude without any spatial extent, such as mass, temperature, specific gravity, number of objects, etc. They are mathematically represented by scalars or real numbers. We call these objects multivectors of grade 0.
2. Second, we consider linelike physical objects having both magnitude and direction, such as displacement, velocity, etc. They are mathematically represented by vectors $\bar{a}, \bar{b}, \ldots$ and graphically by directed line segments. A vector represents the essential abstractions of magnitude and direction without any restriction to its position. We call these linelike objects multivectors of grade 1.
3. Third, we consider planelike physical objects having both magnitude and orientation, such as an oriented flat surface area and the rotation in a plane. It is graphically represented by an oriented parallelogram defined by two vectors $\bar{a}$ and $\bar{b}$ with the head of $\bar{a}$ attached to the tail of $\bar{b}$, and mathematically represented by the bivector $\bar{a} \wedge \bar{b}$, also called the outer product of $\bar{a}$ and $\bar{b}$. A bivector represents the essential abstraction of planar orientation and magnitude without any restriction to the shape of the plane. We call these planelike objects multivectors of grade 2. It is to be noted that the bivector $\bar{a} \wedge \bar{b}$ is different from the usual product $\bar{a} \times \bar{b}$, which is an "axial" vector in the usual vector algebra.
4. Last, we consider volume-like objects having magnitude as well as a choice of handedness, such as an oriented parallelopiped with handedness. It is graphically represented by an oriented parallelopiped defined by three vectors $\bar{a}$, $\bar{b}$, and $\bar{c}$ with the head of $\bar{a}$ attached to the tail of $\bar{b}$ and with the head of $\bar{b}$ attached to the tail of $\bar{c}$, and mathematically represented by trivectors $\bar{a} \wedge \bar{b} \wedge \bar{c}$. The order of vectors in $\bar{a} \wedge \bar{b} \wedge \bar{c}$ determines the handedness and the sign of the oriented parallelopiped. A trivector represents the essential abstraction of volume orientation with handedness and magnitude without any restriction to the shape of the volume. We call these volume-like objects multivectors of grade 3 (see Figure 1.1).

As no mathematical elements with grades greater than 3 can be constructed in three-dimensional Euclidean space, the above-mentioned elements constitute four independent mathematical objects of the geometric algebra for a three-dimensional space. We write a multivector M of any grade as

$$M = |M|(\text{unit of } M),$$

where M is a real number representing the magnitude of M. In geometric algebra for three-dimensional space, unit multivector may be scalar, vector, bivector, or trivector. We take any set of three orthonormal vectors as a basis for vectors. The three mutually orthogonal unit bivectors constructed out of three orthonormal basis vectors are taken as a basis for bivectors. There is only one unit scalar 1. Also, there is only one unit trivector, equal to the product

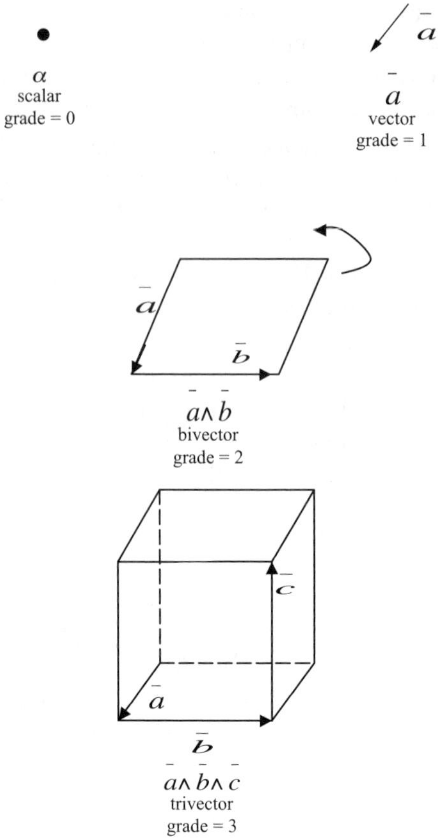

FIGURE 1.1
Four mathematical elements of the geometric algebra for three-dimensional space are represented graphically.

of the three orthonormal vectors considered because there is only one unit volume with the orientation of a given handedness.

A generic multivector M is defined as a linear combination of four linearly independent multivectors of different grades as

$$M = M_0 + M_1 + M_2 + M_3, \tag{1.1}$$

where M_i $(i = 0, 1, 2, 3)$ is a multivector of grade i. The addition of multivectors of different grades may seem absurd at first look. The absurdity disappears because one may justify Equation 1.1 in the abstract Grassmannian way if the indicated relations and operations in mathematics are well defined. For example, a complex number x is defined as a linear combination of a unit scalar 1 and and a unit imaginary j as

$$x = 1x_1 + jx_2. \tag{1.2}$$

Equation 1.2 shows that x has two parts: real and imaginary; they are linearly independent mathematical elements. Likewise, Equation 1.1 shows that M has four parts: scalar (real numbers), vector, bivector, and trivector; all are linearly independent mathematical elements. In the next section we shall show that unit trivector and the unit imaginary have a close resemblance, both being algebraically equal to $\sqrt{-1}$. However, the unit trivector, being a unit volume element with orientation of a given handedness, affords more information, geometrical and physical.

Henceforth we call the multivector of any grade a simple multivector to distinguish it from the generic multivector consisting of four parts: scalar, vector, bivector, and trivector.

1.4 Geometric Algebra as a Symbolic System

Mathematical objects of geometric algebra have one kind of addition rule, different from Gibbs' vector algebra, and one general kind of multiplicative rule, known as the geometric product. The importance of the geometric product of two vectors can be visualized in the fact that all other significant products can be obtained from it. The inner and outer products seem to complement one another by describing independent geometrical relations.

Noting the fact that the inner and outer products of two vectors have opposite symmetries, we define a general kind of product ab (dropping the convention of using overline for vectors) called the geometric product of the vectors a and b, by

$$ab = a \cdot b + a \wedge b. \tag{1.3}$$

By the same mathematical argument given in the previous section we can justify the addition of multivectors of different grades: a scalar (grade 0) and a bivector (grade 2). One can give mathematical meaning to (1.3) by specifying that the addition of scalars and bivectors satisfies the usual commutative and associative rules.

As the inner product obeys commutative rule, we can obtain from (1.3)

$$ba = b \cdot a + b \wedge a = a \cdot b - a \wedge b. \tag{1.4}$$

Here we assume that both the inner and outer products are bilinear in their arguments. So, the geometric product defined by (1.3) is also bilinear in its two arguments.

The geometric product is not generally commutative:

$$ab \neq ba, \tag{1.5}$$

unless $a \wedge b = 0$, for which

$$ab = a \cdot b = b \cdot a = ba, \tag{1.6}$$

nor is it anticommutative:

$$ab \neq -ba, \tag{1.7}$$

unless $a \cdot b = 0$, for which

$$ab = a \wedge b = -b \wedge a = -ba. \tag{1.8}$$

The product ab inherits a geometrical interpretation from those already accorded to the inner and outer products. It is, in fact, an algebraic measure of the relative direction of the vectors a and b as we note that

1. Equation 1.6 implies that the vectors are parallel if and only if their geometric product is commutative.
2. Equation 1.8 implies that the vectors are orthogonal if and only if their geometric product is anticommutative.

As the inner and outer products have opposite symmetries, they can be extracted from (1.3) and (1.4):

$$a \cdot b = (1/2)(ab + ba) \tag{1.9}$$

and

$$a \wedge b = (1/2)(ab - ba). \tag{1.10}$$

Now, instead of regarding (1.3) as the definition of the geometric product ab, we consider it as a fundamental product and take (1.9) and (1.10), respectively, as the definitions of the inner and the outer products of a and b in terms of ab. Thus, in geometric algebra, the composite geometric product is the fundamental algebraic operation with its symmetric and antisymmetric parts being endowed with prime geometrical or physical significance. In this connection one must note that

1. The commutability of the inner product is imparted by the commutability of addition.
2. The anticommutability of the outer product is imparted by the anticommutability of subtraction.

Multiplication of the geometric product ab by a scalar λ gives

$$\lambda(ab) = (\lambda a)b = a(\lambda b), \tag{1.11}$$

which follows from the bilinear property of the geometric product.

The above multiplications are mutually commutative and associative. If the commutative rule is separated from the associative rule by dropping the round brackets in (1.11) we get

$$\lambda a = a\lambda, \tag{1.12}$$

which is the conventional commutative product of a scalar and a vector.

The geometric product obeys the left and right distributive rules:

$$a(b+c) = ab + ac, \tag{1.13}$$

$$(b+c)a = ba + ca \tag{1.14}$$

for any three vectors a, b, and c.

We give the proof of (1.13):

$$\begin{aligned} a(b+c) &= a \cdot (b+c) + a \wedge (b+c) && \text{[definition of geometric product]} \\ &= (a \cdot b + a \cdot c) + (a \wedge b + a \wedge c) && \text{[associative rules for addition]} \\ &= (a \cdot b + a \wedge b) + (a \cdot c + a \wedge c) && \text{[rearrangement of terms]} \\ &= ab + ac && \text{[definition of geometric product].} \end{aligned}$$

In the same way, we can prove (1.14). One must note that the distributive rules (1.13) and (1.14) are independent of one another because the geometric product is, in general, neither commutative nor anticommutative.

In any algebra the associative property is extremely useful in algebraic manipulations. For this purpose we assume that for any three vectors a, b, and c the geometric product is associative:

$$a(bc) = (ab)c = abc. \tag{1.15}$$

Thus, we have ascertained all the basic algebraic properties of the geometric product of vectors including the associative rule.

By exploiting these algebraic properties of the geometric product we will show in what follows that for any three vectors a, b, and c the outer product $a \wedge b \wedge c$ is also associative (see the following Equation 1.22). This can be visualized geometrically by the fact that the mathematical object $a \wedge b \wedge c$ is a volume element with orientation of a given handedness, independent of how the factors of the object are grouped provided the order of the vectors in the product is retained.

One can see easily that the outer product of a vector a and a bivector $A = b \wedge c$ is symmetric:

$$\begin{aligned} a \wedge A &= a \wedge (b \wedge c) = (a \wedge b) \wedge c \\ &= -(b \wedge a) \wedge c = -b \wedge (a \wedge c) \\ &= +b \wedge (c \wedge a) = (b \wedge c) \wedge a \\ &= A \wedge a. \end{aligned} \tag{1.16a}$$

Now we are in a position to extract the inner and outer product of a vector a and a bivector $A = b \wedge c$ from the geometric product aA by using the associative rule (1.15) and noting that $a \cdot A$ and $a \wedge A$ must have opposite symmetries, i.e.,

$$a \wedge A = A \wedge a, \tag{1.16a}$$

$$a \cdot A = -A \cdot a. \tag{1.16b}$$

The anticommutability of the inner product $a \cdot A$ may be seen in the result

$$a \cdot A = a \cdot (b \wedge c) = (a \cdot b)c - (a \cdot c)b$$

calculated later (see the following Equation 1.23).

First we express the geometric product aA as a sum of symmetric and antisymmetric parts:

$$\begin{aligned} aA &= (1/2)(aA + aA) + (1/2)(Aa - Aa) \\ &= (1/2)(aA - Aa) + (1/2)(aA + Aa) \end{aligned} \tag{1.17}$$

and write

$$aA = a \cdot A + a \wedge A, \tag{1.18}$$

where we set in view of (1.16a, b)

$$a \cdot A = (1/2)(aA - Aa) = -A \cdot a \tag{1.19}$$

and

$$a \wedge A = (1/2)(aA + Aa) = A \wedge a. \tag{1.20}$$

All these basic algebraic properties except the associativity of the outer product have already been ascertained. In order to derive the associative rule for the outer product of vectors, we consider the definitions (1.10) and (1.20) and the associative rule (1.15) for the geometric product. Thus we have

$$\begin{aligned} (a \wedge b) \wedge c &= (1/2)[(a \wedge b)c + c(a \wedge b)] \\ &= (1/4)[(ab - ba)c + c(ab - ba)] \\ &= (1/4)[abc - bac + cab - cba]. \end{aligned} \tag{1.21a}$$

Likewise,

$$\begin{aligned} a \wedge (b \wedge c) &= (1/2)[a(b \wedge c) + (b \wedge c)a] \\ &= (1/4)[a(bc - cb) + (bc - cb)a] \\ &= (1/4)[abc - acb + bca - cba]. \end{aligned} \tag{1.21b}$$

From (1.21a, b) we get

$$\begin{aligned} (a \wedge b) \wedge c - a \wedge (b \wedge c) &= (1/4)(cab + acb) - (1/4)(bac + bca) \\ &= (1/4)(ca + ac)b - (1/4)b(ac + ca) \\ &= (1/2)(c \cdot a)b - (1/2)b(a \cdot c) \\ &= (1/2)(a \cdot c)b - (1/2)(a \cdot c)b \\ &= 0. \end{aligned}$$

Thus we have

$$(a \wedge b) \wedge c = a \wedge (b \wedge c), \tag{1.22}$$

which gives the associative rule for the outer product of vectors.

The symmetric part $a \wedge A$ of the geometric product aA in (1.18) is identified with the outer product of a vector and a bivector, which is, in fact, a trivector $a \wedge b \wedge c$, a multivector of grade 3.

The antisymmetric part $a \cdot A$ of the geometric product aA in (1.18) is identified with the inner product of a vector and a bivector, which may be regarded as a generalization of the inner product of vectors. In order to understand the significance of the quantity $a \cdot A$, one must expand it explicitly in terms of the inner product of two vectors to exibit its grade for ascertaining the mathematical object it represents.

By using the definitions (1.9), (1.10), (1.19) and the associative rule (1.15) for the geometric product, one can write, taking $A = b \wedge c$:

$$\begin{aligned}
a \cdot A &= (1/2)[aA - Aa] = (1/2)[a(b \wedge c) - (b \wedge c)a] \\
&= (1/4)[a(bc - cb) - (bc - cb)a] \\
&= (1/4)[a(bc) - a(cb) - (bc - cb)a] \\
&= (1/4)[(ab)c - (ac)b - (bc - cb)a] \\
&= (1/4)[(2a \cdot b - ba)c - (2a \cdot c - ca)b - (bc - cb)a] \\
&\quad \{\textit{remember } ab = (2a \cdot b - ba), \textit{etc.}\} \\
&= (1/4)[(2a \cdot b)c - (ba)c - (2a \cdot c)b + (ca)b - (bc - cb)a] \\
&= (1/4)[(2a \cdot b)c - b(ac) - (2a \cdot c)b + c(ab) - (bc - cb)a] \\
&= (1/4)[(2a \cdot b)c - b(2a \cdot c - ca) - (2a \cdot c)b \\
&\quad + c(2a \cdot b - ba) - (bc)a + (cb)a] \\
&= (1/4)[(2a \cdot b)c - (2a \cdot c)b + b(ca) - (2a \cdot c)b \\
&\quad + (2a \cdot b)c - c(ba) - b(ca) + c(ba)] \\
&= (a \cdot b)c - (a \cdot c)b.
\end{aligned}$$

Thus we have

$$a \cdot A = a \cdot (b \wedge c) = (a \cdot b)c - (a \cdot c)b. \tag{1.23}$$

This shows that the inner product of a vector and a bivector is anticommutative and represents a vector. In the derivation of the result (1.23) we first write the expression simply in terms of geometric products and then repeatedly use the associative rule for the geometric product and the inner product of two vectors, written as $ab = 2a \cdot b - ba$. This demonstrates that the composite geometric product with its associative property is a fundamental algebraic operation.

Equations 1.9, 1.10 and 1.19, 1.20 demonstrate the general rules for the inner and outer products, which may be stated as

1. The inner product by a vector lowers the grade of any simple multivector by one.
2. The outer product by a vector raises the grade of any simple multivector by one.

One may note the following pattern of symmetry for the outer product of a vector a and multivectors of different grades. It is antisymmetric for any vector b(multivector of grade 1):

$$a \wedge b = -b \wedge a, \tag{1.24}$$

and symmetric for any bivector (multivector of grade 2) $A = b \wedge c$:

$$a \wedge A = A \wedge a, \tag{1.25}$$

which shows that symmetry alternates with grade. The above symmetry may be generalized by the rule

$$a \wedge M = (-1)^g M \wedge a, \tag{1.26}$$

where a is any vector and M is any multivector of grade "g".

Also noting that the inner and outer product of a vector and any multivector of grade g must have opposite symmetries, and taking account of the symmetry for the outer products as given by (1.26), we can express the symmetry for the inner products by the rule:

$$a \cdot M = -(-1)^g M \cdot a. \tag{1.27}$$

This can also be obtained as the generalization of the results (1.9) and (1.23).

1.5 Geometric Algebra as an Axiomatic System (Axiom A . . .)

In Section 1.4 we have introduced geometric algebra for three-dimensional space as a symbolic system that includes graded multivectors $M_i (i = 0, 1, 2, 3)$, called simple multivectors (scalar, vector, bivector, and trivector) to represent the directional properties of points, lines, planes, and space (volume). Because the graded multivectors M_i are linearly independent mathematical objects, we define a generic multivector M, a mathematical object of "mixed" grades, to be a linear combination of them as

$$M = M_0 + M_1 + M_2 + M_3. \tag{1.28}$$

Any element of geometric algebra can be expressed in the form (1.28). In this type of addition, multivectors of different grades do not mix; they are simply collected as separate parts under one heading called multivector. As in the addition of real and imaginary numbers, numbers of different types are collected as separate parts under the name of complex numbers.

We note in passing that the geometric product of vectors has, except for commutativity, the same algebraic properties as the scalar multiplication of vectors and bivectors. In particular, both products are associative as well as distributive with respect to addition.

Now, in conformity with the development of geometric algebra for three-dimensional space as a symbolic system, we develop the geometric algebra for the space of an arbitrary dimension by introducing the following axioms and definitions.

We denote by $\mathcal{G}$ the geometric algebra for a space of an arbitrary dimension, and $A, B, C, \ldots$ are multivectors belonging to $\mathcal{G}$.

Axiom 1 : $\mathcal{G}$ is closed under the addition of any two multivectors belonging to $\mathcal{G}$, i.e., for any two multivectors

$$A, B, \in \mathcal{G}$$

there exists a unique multivector $C \in \mathcal{G}$, such that

$$A + B = C \tag{A.1}$$

Axiom 2 : $\mathcal{G}$ is closed under the multiplication (geometric) of any two multivectors $A, B \in \mathcal{G}$, i.e., there exists a unique multivector $C \in \mathcal{G}$, such that

$$AB = C. \tag{A.2}$$

Axiom 3: Addition of multivectors $A, B \in \mathcal{G}$ is commutative, i.e.,

$$A + B = B + A. \tag{A.3}$$

Axiom 4: Addition is associative, i.e., for any three multivectors A, B, and $C \in \mathcal{G}$, we have

$$(A + B) + C = A + (B + C). \tag{A.4}$$

Axiom 5: The geometric product of multivectors $\in \mathcal{G}$ is associative, i.e., for any three multivectors $A, B, C \in \mathcal{G}$,

$$(AB)C = A(BC). \tag{A.5}$$

Axiom 6: The geometric product of multivectors $\in \mathcal{G}$ obeys the left and right distributive rules with respect to addition, i.e., for any three multivectors $A, B, C \in \mathcal{G}$,

$$A(B + C) = AB + AC. \tag{A.6}$$

$$(B + C)A = BA + CA. \tag{A.7}$$

Note that the distributive rules (A.6) and (A.7) are independent of one another because neither commutability nor anticommutability of the geometric product of multivectors is axiomatized.

Axiom 7: There exists a unique multivector $0 \in \mathcal{G}$, called the additive identity, such that

$$A + 0 = A = 0 + A. \tag{A.8}$$

Axiom 8: There exists a unique multivector $I \in \mathcal{G}$, called the multiplicative identity, such that

$$IA = A. \tag{A.9}$$

Axiom 9: Every multivector $A \in \mathcal{G}$ has a unique multivector $-A \in \mathcal{G}$, called the additive inverse, such that

$$A + (-A) = 0 = (-A) + A. \tag{A.10}$$

Axiom 10: The set of scalars in the algebra are real numbers.

Axiom 11: The multiplication of a multivector $A \in \mathcal{G}$ by a scalar λ is commutative:

$$\lambda A = A\lambda. \tag{A.11}$$

Axiom 12: The square of any non-zero vector a is a unique position scalar $|a|^2$:

$$a^2 = |a|^2 > 0. \tag{A.12}$$

Axiom 13: For every non-zero vector $a \in \mathcal{G}$ there exists a unique vector $a^{-1} \in \mathcal{G}$, called the multiplicative inverse, such that

$$aa^{-1} = I = a^{-1}a, \tag{A.13}$$

where

$$a^{-1} = a/a^2. \tag{A.14}$$

Axiom 14: For every vector a and a multivector A_r of grade r in an r-dimensional space,

$$a \wedge A_r = 0. \tag{A.15}$$

The left-hand member is a multivector of grade $r + 1$(see rules 1 and 2 of Section 1.4); this axiom is necessary because r-dimensional space does not allow any multivector with grade greater than r.

On the other hand, from the geometrical point of view, we can say that if $a \wedge A_r$ is $\neq 0$, we will be in a vector space that has a dimension not lower than $r + 1$ because, being $a \wedge A_r \neq 0$, $a \wedge A_r$ is an $(r + 1)$-multivector that can be only in an $r + 1$ space. Then, in an r-dimensional space, the outer product of a vector by any multivector of grade r must be 0, i.e., $a \wedge A_r = 0$. We can say then in this manner: no mathematical objects with grade greater than r can be constructed in an r-dimensional space (see also the proposition given following Figure 1.1, which refers to a three-dimensional Euclidean space, where it is said that no mathematical elements with grade greater than 3 can be constructed). So, for example, in three-dimensional physical space we have

$$a \wedge A_3 = 0. \tag{1.29}$$

Next, we give some definitions. For a vector a and any multivector A_k of grade k we define the inner product by

$$a \cdot A_k = (1/2)(aA_k - (-1)^k A_k a) = -(-1)^k A_k \cdot a \tag{1.30}$$

and the outer product by

$$a \wedge A_k = (1/2)(aA_k + (-1)^k A_k a) = (-1)^k A_k \wedge a. \tag{1.31}$$

Adding (1.30) and (1.31) we have the geometric product aA_k as

$$aA_k = a \cdot A_k + a \wedge A_k \tag{1.32}$$

Note:

1. (1.30) includes (1.9) and (1.19) as special cases.
2. (1.31) includes (1.10) and (1.20) as special cases.
3. (1.32) includes (1.3) and (1.18) as special cases.

From the definitions (1.30) and (1.31) we adopt the following rules in accordance with the rules depicted in Section 1.4 for lowering and raising the grades of any multivector by its inner and outer products with a vector:

1. $a \cdot A_k$ is a multivector of grade $k - 1$. (1.33)
2. $a \wedge A_k$ is a multivector of grade $k + 1$. (1.34)

In particular, the inner product of a multivector λ of grade 0 by a vector, following the rule 1, (1.33) has a grade 0–1, and then it is without any meaning (and thus is not an element of geometric algebra).

By using the definition (1.31) one can write eq.(1.29) as

$$aA_3 = A_3 a. \tag{1.35}$$

We have given this example contained in Equation 1.35 because it will be useful when we will consider the Pauli algebra.

Notice that axioms A.1 to A.9 implicitly define the operations of addition, subtraction, and multiplication of mathematical elements of the geometric algebra. Except for the commutative law for multiplication, they are identical with the axioms of scalar algebra.

We will give here the derivation of the associative rule for the outer product of vectors:

$$a \wedge (b \wedge c) = (a \wedge b) \wedge c. \tag{1.36}$$

We begin with the associative rule for the geometric product:

$$a(bc) = (ab)c \tag{1.37}$$

$$a(b \cdot c + b \wedge c) = (a \cdot b + a \wedge b)c \qquad \text{[definition]} \tag{1.38}$$

$$a(b \cdot c) + a(b \wedge c) = (a \cdot b)c + (a \wedge b)c \qquad \text{[distributive law]} \tag{1.39}$$

$$a(b \cdot c) + a \cdot (b \wedge c) + a \wedge (b \wedge c)$$
$$= (a \cdot b)c + (a \wedge b) \cdot c + (a \wedge b) \wedge c \qquad \text{[definition]} \tag{1.40}$$

By using rule (1.33) we identify the terms $a(b \cdot c)$, $a \cdot (b \wedge c)$, $(a \cdot b)c$, and $(a \wedge b) \cdot c$ as vectors. Likewise, by using rule (1.34), the terms $a \wedge (b \wedge c)$ and $(a \wedge b) \wedge c$ are identified as trivectors. By equating the trivector parts from both sides, we get the associative rule for the outer product:

$$a \wedge (b \wedge c) = (a \wedge b) \wedge c. \tag{1.41}$$

Also, by equating the vector parts, we get an algebraic identity:

$$a(b \cdot c) + a \cdot (b \wedge c) = (a \cdot b)c + (a \wedge b) \cdot c. \tag{1.42}$$

Now we derive the distributive rules for the inner and outer products. By using the left distributive rule (A.6) of the geometric product for a vector a and r-grade multivectors B_r and C_r we get

$$a(B_r + C_r) = aB_r + aC_r \tag{1.43}$$

$$a \cdot (B_r + C_r) + a \wedge (B_r + C_r) \tag{1.44}$$
$$= a \cdot B_r + a \wedge B_r + a \cdot C_r + a \wedge C_r \qquad \text{[using (1.32)].}$$

By taking account of rules (1.33) and (1.34), we equate the multivectors of like grades from both sides and get the left distributive rules for the inner and outer products:

$$a \cdot (B_r + C_r) = a \cdot B_r + a \cdot C_r \tag{1.45}$$

and

$$a \wedge (B_r + C_r) = a \wedge B_r + a \wedge C_r. \tag{1.46}$$

Similarly, by using the right distributive rule (A.7) we get the right distributive rules for the inner and outer product:

$$(B_r + C_r) \cdot a = B_r \cdot a + C_r \cdot a \tag{1.47}$$

and

$$(B_r + C_r) \wedge a = B_r \wedge a + C_r \wedge a. \tag{1.48}$$

A consequence of rule (1.33) is that if we take the inner product of a vector a by a multivector λ of grade 0, we will find a multivector of grade -1. As this is impossible, this kind of inner product is meaningless. Again, a consequence of rule (1.34) is that if we take the outer product of a vector a by a multivector λ of grade 0, we have a multivector of grade 1; in that case, $a\lambda = \lambda a$ is the same as the conventional product of a scalar and a vector.

1.6 Some Essential Formulas and Definitions

According to definition (1.28) of Section 1.5, every multivector A in three-dimensional Euclidean space can be expressed linearly in terms of graded multivectors $A_k (k = 0, 1, 2, 3)$ as

$$A = A_0 + A_1 + A_2 + A_3. \tag{1.49}$$

Multivector A is said to be even if it contains only the even-graded multivector parts A_0 and A_2, and odd if it contains only the odd-graded multivector parts A_1 and A_3.

Denoting the even and odd multivector parts by A_+ and A_-, respectively, we see that

$$A = A_+ + A_- \tag{1.50}$$

where

$$A_+ = A_0 + A_2 \tag{1.51}$$

and

$$A_- = A_1 + A_3. \tag{1.52}$$

It is easy to show that even multivectors form an algebra by themselves, which is a subalgebra of the full geometric algebra, but odd multivectors do not.

In any multivector containing products of different kinds, we perform the operations of multiplication in the following order: outer product, inner product and, last, geometric product. This convention of preference order of performing multiplications operations removes the ubiquitous use of parenthesis. The following are examples:

$$A \wedge BC = (A \wedge B)C \neq A \wedge (BC) \tag{1.53}$$

$$A \cdot BC = (A \cdot B)C \neq A \cdot (BC) \tag{1.54}$$

$$A \cdot B \wedge C = A \cdot (B \wedge C) \neq (A \cdot B) \wedge C, \tag{1.55}$$

where A, B, and C are graded multivectors.

Now we present the following reduction formula and its expanded form without proof. Their proofs are presented in Reference [3].

1. The general reduction formula:

$$a \cdot a_1 \wedge C_{r-1} = a \cdot a_1 C_{r-1} - a_1 \wedge (a \cdot C_{r-1}), \tag{1.56}$$

   where a and a_1 are vectors, and C_{r-1} is a multivector of grade $r-1$. This is valid for any positive integral value of $r \geq 3$.
2. By iterating $a_1 \wedge (a \cdot C_{r-1})$, we write it in the expanding form:

$$\begin{aligned} & a_1 \wedge (a \cdot a_2 \wedge a_3 \wedge \cdots \wedge a_r) \\ & = a \cdot a_2 a_1 \wedge a_3 \wedge \cdots \wedge a_r -_r a \cdot a_3 a_1 \wedge a_2 \wedge a_4 \wedge \cdots \wedge a \\ & \quad + \cdots + (-1)^r a \cdot a_r a_1 \wedge a_2 \wedge \cdots \wedge a_{r-1}. \end{aligned} \tag{1.57}$$

This is valid for any positive integral value of $r \geq 3$. Particular case: if $r = 3$, we get

$$a_1 \wedge (a \cdot a_2 \wedge a_3) = a \cdot a_2 a_1 \wedge a_3 - a \cdot a_3 a_1 \wedge a_2. \tag{1.58}$$

This can be directly obtained by using the formula

$$a \cdot a_2 \wedge a_3 = (a \cdot a_2) a_3 - (a \cdot a_3) a_2. \tag{1.59}$$

In view of (1.57) the general reduction formula (1.56) can be expressed in the following expanded form:

$$\begin{aligned} a \cdot (a_1 \wedge a_2 \wedge \cdots \wedge a_r) &= \sum_{s=1}^{r} (-1)^{s+1} a \cdot a_s a_1 \wedge a_2 \wedge \breve{a}_s \cdots \wedge a_r \\ &= a \cdot a_1 a_2 \wedge a_3 \wedge \cdots \wedge a_r - a \cdot a_2 a_1 \wedge a_3 \wedge \cdots \wedge a_r + \cdots \\ &\quad + (-1)^{r+1} a \cdot a_r a_1 \wedge a_2 \wedge \cdots \wedge a_{r-1}, \end{aligned} \tag{1.60}$$

where the invested circumflex in the product

$$a \cdot a_s a_1 \wedge a_2 \wedge \cdots \breve{a}_s \cdots \wedge a_r \tag{1.61}$$

means that the a_s factor is to be omitted.

Equation 1.60 determines the inner product of a vector a and an r-graded multivector $A_r = a_1 \wedge a_2 \wedge \cdots \wedge a_r$. The general reduction formula (1.56) has been referred to as the Laplace expansion of the inner product by Hestenes [3].

Now, we generalize the reduction formulas (1.56) and (1.60) by taking a multivector of any grade in place of vector a. For the sake of convenience, we write $(A)_r$ to denote the r-graded multivector part of a multivector A.

For any r-graded multivector A_r and s-graded multivector B_s, we define the inner product $A_r \cdot B_s$ by

$$A_r \cdot B_s \equiv (A_r B_s)_{|r-s|} \tag{1.62}$$

and the outer product $A_r \wedge B_s$ by

$$A_r \wedge B_s \equiv (A_r B_s)_{r+s}. \tag{1.63}$$

The inner product produces an $|r-s|$-graded multivector, whereas the outer product produces an $(r+s)$-graded multivector.

In the following text we deduce three important formulas and the associative rule for the generalized outer product.

We begin with the associative rule for the geometric product of an r-graded multivector A_r, an s-graded multivector C_s, $(0 < r < s)$, and a vector b:

$$(A_r b)C_s = A_r(bC_s). \tag{1.64}$$

$$(A_r \cdot b + A_r \wedge b)C_s = A_r(b \cdot C_s + b \wedge C_s) \quad \text{[definition]}. \tag{1.65}$$

$$(A_r \cdot b)C_s + (A_r \wedge b)C_s = A_r(b \cdot C_s) + A_r(b \wedge C_s) \quad \text{[definition]}. \tag{1.66}$$

$$(A_r \cdot b) \cdot C_s + (A_r \cdot b) \wedge C_s + (A_r \wedge b) \cdot C_s + (A_r \wedge b) \wedge C_s. \tag{1.67}$$

$$= A_r \cdot (b \cdot C_s) + A_r \wedge (b \cdot C_s) + A_r \cdot (b \wedge C_s) + A_r \wedge (b \wedge C_s) \quad \text{[definition]}. \tag{1.68}$$

Now we equate like graded multivector parts from both sides of Equation 1.68.

1. Equating the $(s+r+1)$-graded multivector parts we get the associative rule for the generalized outer product:

$$(A_r \wedge b) \wedge C_s = A_r \wedge (b \wedge C_s). \tag{1.69}$$

2. Equating the $(s+r-1)$-graded multivector parts we get

$$(A_r \cdot b) \wedge C_s = A_r \wedge (b \cdot C_s). \tag{1.70}$$

3. Equating the $(s-r+1)$-graded multivector parts we get

$$(A_r \cdot b) \cdot C_s = A_r \cdot (b \wedge C_s). \tag{1.71}$$

4. Finally, equating the $(s-r-1)$-graded multivector parts, we get

$$(A_r \wedge b) \cdot C_s = A_r \cdot (b \cdot C_s). \tag{1.72}$$

The factor $(b \cdot C_s)$ on the right-hand side of (1.70) or (1.72) can be reduced to the expanded form of the general reduction formula (1.60) if C_s is expressed as an outer product of s vectors.

We have some particular cases:

1. The equation relating the bivector parts:

$$(A_2 \cdot b) \cdot C_3 = A_2 \cdot (b \wedge C_3) \tag{1.73}$$

is a particular case of (1.71).

2. The equations relating the scalar parts

$$(a \wedge b) \cdot C_2 = a \cdot (b \cdot C_2) \tag{1.74}$$

and

$$(A_2 \wedge b) \cdot C_3 = A_2 \cdot (b \cdot C_3) \tag{1.75}$$

and that relating the vector parts

$$(a \wedge b) \cdot C_3 = a \cdot (b \cdot C_3) \tag{1.76}$$

are particular cases of (1.72). Equation 1.74 is useful in three-dimensional Euclidean space, whereas Equation 1.73, Equation 1.75, and Equation 1.76 are useful in four-dimensional space-time in simplifying algebraic expressions.

References

1. D. Hestenes, *Space-Time Algebra* (Gordon and Breach, New York, 1966).
2. D. Hestenes and G. Sobczyck, *Clifford Algebra to Geometric Calculus* (Reidel, Boston, 1984).
3. D. Hestenes, *New Foundations for Classical Mechanics* (Reidel, Boston, 1986).
4. B. K. Datta, V. de Sabbata, and R. Datta, *Sci. Cult.*, 65, 64–67 (1999).
5. R. Penrose, *The Emperor's New Mind*, Oxford University Press, Oxford, U.K., (1989).
6. B. K. Datta and R. Datta *J. Birla Planet.*, Vol.8. No. 2, 30–32.
7. L. Russo, *La Rivoluzione Dimenticata*, second edition, Feltrinelli Editore, Milano, Italia (1997).
8. T. G. Vold, *Am. J. Phys.* 61, 491–504 (1993).

2

Multivectors

2.1 Geometric Product of Two Bivectors A and B

Expressing bivector A as a product of orthogonal vectors:

$$A = a \wedge b = ab, \tag{2.1}$$

we can write

$$\begin{aligned} AB &= abB = a(bB) = a(b \cdot B + b \wedge B) \\ &= a(b \cdot B) + a(b \wedge B) \\ &= a \cdot (b \cdot B) + a \wedge (b \cdot B) + a \cdot (b \wedge B) + a \wedge (b \wedge B) \\ &= (a \wedge b) \cdot B + a \wedge (b \cdot B) + a \cdot (b \wedge B) + (a \wedge b) \wedge B \\ &\qquad\qquad\qquad\qquad\qquad\qquad\qquad\qquad \text{[by using (1.74)]} \\ &= A \cdot B + [a \wedge (b \cdot B) + a \cdot (b \wedge B)] + A \wedge B \\ &= (AB)_o + (AB)_2 + (AB)_4, \end{aligned} \tag{2.2}$$

where

$$(AB)_o = A \cdot B = a \cdot (b \cdot B) = (BA)_o, \tag{2.3}$$

$$(AB)_2 = a \wedge (b \cdot B) + a \cdot (b \wedge B) = -(BA)_2, \tag{2.4}$$

$$(AB)_4 = A \wedge B = a \wedge b \wedge B = (BA)_4. \tag{2.5}$$

We decompose the geometric product AB into symmetric and antisymmetric part [1]:

$$AB = (1/2)(AB + BA) + (1/2)(AB - BA) \tag{2.6}$$

Comparing (2.2) with (2.6) and noting the relations (2.3)–(2.5), we establish that

$$(AB)_o + (AB)_4 = (1/2)(AB + BA) = (BA)_o + (BA)_4$$

and

$$(AB)_2 = (1/2)(AB - BA) = -(BA)_2.$$

Stated more explicitly, we can write

$$A \cdot B + A \wedge B = (1/2)(AB + BA) = B \cdot A + B \wedge A, \tag{2.7}$$

and

$$a \wedge (b \cdot B) + a \cdot (b \wedge B) = (1/2)(AB - BA). \tag{2.8}$$

The expression $(1/2)(AB - BA)$ is called the commutator or commutator product of A and B. In three-dimensional space

$$A \wedge B = 0$$

Equation 2.2 and Equation 2.7 become

$$AB = (AB)_0 + (AB)_2 = A \cdot B + a \wedge (b \cdot B) + a \cdot (b \wedge B) \tag{2.9}$$

and

$$A \cdot B = (1/2)(AB + BA) = B \cdot A. \tag{2.10}$$

The geometric product of bivectors can be generalized to the geometric product of multivectors of any grades, $A_r B_s$. For the geometric product

$$A_r B_s = a_1 a_2 \ldots a_r B_s, \quad (r \leq s),$$

the term of the lowest grade will be

$$A_r \cdot B_s = (A_r B_s)_{s-r}. \tag{2.11}$$

Corollary: The geometric product $A_r B_s$ can have a nonzero scalar part

$$A_r \cdot B_s \quad \text{if } r = s.$$

Factorization: In geometric algebra there is a type of factorizing an r-graded multivector into an outer product of vectors.

We see how a nonzero bivector B can be factorized into an outer product of vectors.

We take a unit vector a and a nonzero bivector B such that

$$a \wedge B = 0 \tag{2.12}$$

This condition implies that the unit vector a lies in the plane of the bivector B. Then we can write

$$aB = a \cdot B \equiv b \tag{2.13}$$

This defines a unique vector b. We can solve Equation 2.13 for B in terms of a and b. For this purpose we multiply (2.13) on the left by $a^{-1} = a$,

$$\begin{aligned} a^{-1}aB &= a^{-1}b, \\ (a^{-1}a)B &= ab, \\ B &= ab. \end{aligned} \tag{2.14}$$

From 2.12 and 2.13 we get

$$a \cdot b = a \cdot (a \cdot B)$$
$$= (a \wedge a) \cdot B \qquad \text{[using (1.74)]}.$$

Thus we have

$$a \cdot b = 0. \tag{2.15}$$

So, b is orthogonal to a. Thus, from (2.14) and (2.15) we obtain

$$B = ab = a \wedge b. \tag{2.16}$$

Equation 2.16 gives us a factorization of the bivector B into an outer product of orthogonal vectors with the condition (2.12) that the unit vector a be a factor of B. Equation 2.13 tells us that b is a unique factor of B orthogonal to a.

2.2 Operation of Reversion

In geometric algebra we introduce one kind of conjugation called reversion [1]. The reverse of any multivector A, denoted by Ã, is defined to be the expression obtained from A by reversing the order of all vector factors in all simple multivectors making up A.

The reverse of a bivector $B = a \wedge b$ is given by

$$\widehat{B\dagger} = (a \wedge b)\tilde{} = b \wedge a = -a \wedge b = -B. \tag{2.17}$$

The reverse of a trivector $T = a \wedge b \wedge c$ is given by

$$\tilde{T} = (a \wedge b \wedge c)^{=} c \wedge b \wedge a = -(b \wedge c) \wedge a$$
$$= b \wedge (a \wedge c) = -a \wedge (b \wedge c) = -T. \tag{2.18}$$

Thus reversion changes the signs of bivectors and trivectors. Scalars and vectors, on the other hand, remain unchanged. So, the reverse of a general multivector A in the expanded form is given by

$$\tilde{A} = (A_0 + A_1 + A_2 + A_3)\tilde{} = A_0 + A_1 - A_2 - A_3. \tag{2.19}$$

It follows from the definition that the reverse of a geometric product of multivectors is the reverse of the reversed multivectors:

$$(AB \ldots CD)\tilde{} = \tilde{D}\tilde{C} \ldots \tilde{B}\tilde{A}. \tag{2.20}$$

In particular, the reverse of a geometric product of vectors is given by

$$(a_1 a_2 \ldots a_r)\tilde{} = a_r \ldots a_2 a_1. \tag{2.21}$$

Reversion corresponds to Hermitian conjugation in matrix algebra and is very useful in the problems of rotation where pairs of bivectors, one the reverse of the other, occur in the usual way.

2.3 Magnitude of a Multivector

To every multivector A there corresponds a unique scalar $|A|$, called the magnitude or modulus of A, defined by the equation

$$|A| = (\tilde{A}A)_0^{1/2}. \tag{2.22}$$

To prove the existence of (2.22) we have to show that

$$|A|^2 = (A \uparrow A)_0 \geq 0, \tag{2.23}$$

where $|A| = 0$ if and only if $A = 0$.

Proof of the existence theorem (2.23):
First we observe that

$$|a_1 \ldots a_r|^2 = (a_1 \ldots a_r)\tilde{}(a_1 \ldots a_r) = |a_1|^2 \ldots |a_r|^2 \geq 0 \tag{2.24}$$
$$\text{if } (a_1 \ldots a_r) \neq 0.$$

If the vectors are orthogonal, they are factors of an r-graded multivector A_r:

$$A_r = a_1 a_2 \ldots a_r = a_1 \wedge a_2 \wedge \ldots \wedge a_r. \tag{2.25}$$

Then it follows that for any multivector of grade r

$$|\ A_r\ |^2 \geq 0, \quad \text{if } A_r \neq 0. \tag{2.26}$$

In the expansion of the scalar part of the product (ÃA), the cross terms of products of multivectors of different grades should be omitted as they have no scalar parts. Thus we have

$$|A|^2 = |\tilde{A}A|_0 = |A_0|^2 + |A_1|^2 + \cdots + |A_r|^2 \geq 0. \tag{2.27}$$

Hence existence of theorem (2.23) is proved.

2.4 Directions and Projections

In geometric algebra the notion of "direction" is given a precise mathematical representation by a "unit vector," so the unit vectors themselves are referred to as *directions*.

Consider the geometric product of two vectors a and b:

$$ab = a \cdot b + a \wedge b. \tag{2.28}$$

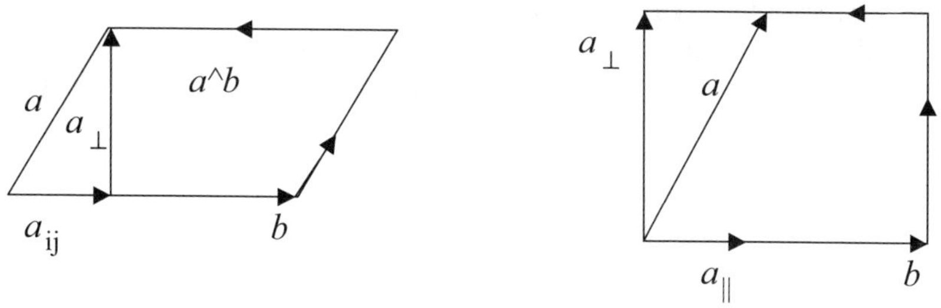

FIGURE 2.1
$a_{||}$ is collinear with b, and $a_{\perp}$ is perpendicular to b.

The interpretations associated with the inner and outer products $a \cdot b$ and $a \wedge b$ imply that

1. Vectors a and b are *collinear* if and only if $ab = ba$.
2. They are *orthogonal* if and only if $ab = -ba$.

From the above implications the geometric product ab may be considered to be an algebraic measure of the relative "directions" of vectors a and b somewhere between these two extremes.

The resolved parts of vector a along and perpendicular to vector b can be obtained by right multiplication of (2.28) by b^{-1}as shown by [1]

$$abb^{-1} = a \cdot bb^{-1} + a \wedge bb^{-1}$$

or

$$a = a \cdot bb^{-1} + a \wedge bb^{-1}. \tag{2.29}$$

The parts $a \cdot bb^{-1}$ and $a \wedge bb^{-1}$ are, respectively, the resolved parts of the vector a along and perpendicular to the vector b as shown in Figure 2.1.

Now, setting

$$a_{||} = a \cdot bb^{-1}, \tag{2.30a}$$

$$a_{\perp} = a \wedge bb^{-1}, \tag{2.30b}$$

we can write Equation 2.29 as

$$a = a_{||} + a_{\perp}. \tag{2.31}$$

(2.30a, b) can be expressed by the equations

$$a_{||}b = a \cdot b = ba_{||}, \tag{2.32a}$$

$$a_{\perp}b = a \wedge b = -ba_{\perp}. \tag{2.32b}$$

(2.32b) expresses the directed area $a \wedge b$ as the product of the *altitude* $a_\perp$ and *base* b of the (a, b)-parallelogram.

Similarly, the resolved parts $b_{||}$ and $b_\perp$, respectively, of vector b along and perpendicular to vector a can be obtained by left multiplication of (2.28) by a^{-1} as

$$a^{-1}ab = a^{-1}a \cdot b + a^{-1}a \wedge b$$

that is,

$$b = b_{||} + b_\perp, \tag{2.33}$$

where

$$b_{||} = a^{-1}a \cdot b, \tag{2.34a}$$

$$b_\perp = a^{-1}a \wedge b. \tag{2.34b}$$

(2.34a, b) can be expressed by the equations

$$ab_{||} = a \cdot b = b_{||}a, \tag{2.35a}$$

$$ab_\perp = a \wedge b = -b_\perp a. \tag{2.35b}$$

We have seen earlier that a bivector B determines a two-dimensional vector space called B-space. The relative direction of B and some vector a is completely characterized by the geometric product

$$aB = a \cdot B + a \wedge B. \tag{2.36}$$

As in the earlier case, vector a is *uniquely* resolved into a vector $a_{||}$ in the B-space and a vector $a_\perp$ orthogonal to the B-space as given by

$$a = a_{||} + a_\perp. \tag{2.37}$$

where

$$a_{||} = a \cdot BB^{-1}, \tag{2.38a}$$

$$a_\perp = a \wedge BB^{-1}. \tag{2.38b}$$

The relations (2.38a, b) are represented in Figure 2.2.

By using Equation 2.30 and Equation 2.31, the Equations 2.38a, b can be expressed by the equations

$$a_{||}B = a \cdot B = -Ba_{||}, \tag{2.39a}$$

$$a_\perp B = a \wedge B = Ba_\perp. \tag{2.39b}$$

The above equations imply that a vector is in the B-space (plane) if and only if it anticommutes with B, and it is orthogonal to the B-space (plane) if and only if it commutes with B.

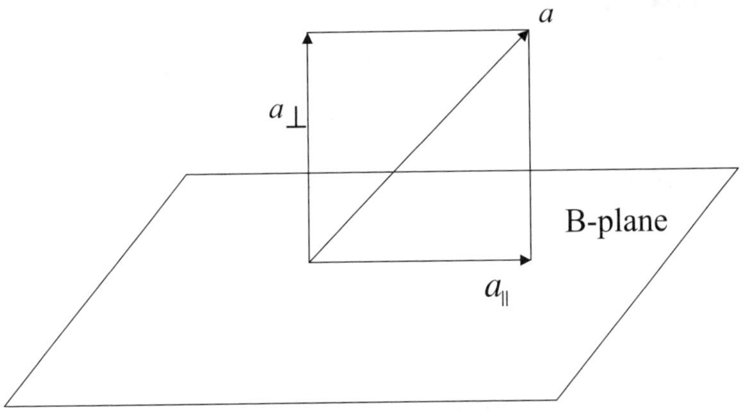

FIGURE 2.2
Projection and rejection of vector by a bivector B.

Next, we generalize the above case for a multivector M of an arbitrary grade k, which determines the ak-dimensional vector space called M-space. The relative direction of M and some vector a is completely characterized by the geometric product

$$aM = a \cdot M + a \wedge M. \tag{2.40}$$

As in the above case, the vector a is *uniquely* resolved into a vector $a_{\shortparallel}$ in the M-space, and a vector $a_\perp$ orthogonal to the M-space as given by

$$a = a_{||} + a_\perp, \tag{2.41}$$

where

$$a_{||} = a \cdot MM^{-1}, \tag{2.42a}$$

$$a_\perp = a \wedge MM^{-1}. \tag{2.42b}$$

By using Equation 2.30 and Equation 2.31, Equations (2.42a, b) can be expressed by the equations

$$a_{||}M = a \cdot M = (-1)^{k+1}Ma_{||}, \tag{2.43a}$$

$$a_\perp M = a \wedge M = (-1)^k Ma_\perp. \tag{2.43b}$$

The above equations are the generalization of Equations (2.39a, b). The vector $a_{||}$ determined by Equation 2.42a is called the *projection* of vector a into the M-space, whereas $a_\perp$ determined by Equation (2.42b) is called the *rejection* of vector a from the M-space.

2.5 Angles and Exponential Functions (as Operators)

An angle is a relation between two directions. Now, following Hestenes [1], we give a precise mathematical expression for this relation.

If ϑ is the angle between the direction a and b (unit vectors), then the cosine and sine of the angle ϑ respectively, are defined by the equations

$$a \cdot b = \cos\vartheta \tag{2.44a}$$

$$a \wedge b = i \sin\vartheta \tag{2.44b}$$

where i is the unit pseudoscalar of the $a \wedge b$-plane.

Now we can write, in view of (2.44 a, b),

$$z = ab = a \cdot b + a \wedge b = \cos\vartheta + i \sin\vartheta$$

giving

$$z = ab = e^{i\vartheta}, \tag{2.45}$$

where

$$e^{i\vartheta} = \cos\vartheta + i \sin\vartheta, \tag{2.46}$$

$$|z| = 1. \tag{2.47}$$

This shows that Equations (2.44.a, b) are just parts of the single fundamental equation (2.45), which indicates that $e^{i\vartheta}$ is a spinor of the i-plane. Equation 2.46 may be regarded as a definition of the exponential function $\exp(i\vartheta)$ from the operational point of view.

We use the radian measure of the angle ϑ almost exclusively because the degree measure is not compatible with the fundamental definition (algebraic) of the exponential function we discuss in the next section. Moreover, in what follows, we represent angles by bivectors, where the "areal measure" is not compatible with degree measure. The angle ϑ in Equation 2.45 is interpreted as "radian measure" of the angle from a to b. This means that the numerical magnitude of ϑ is equal to the length of the arc on the unit circle from a to b as indicated in Figure 2.3. As an angle is a relation between two directions that determine a plane, we represent angle by a bivector. So the angle from a to b is represented by the bivector ϕ given by

$$\phi = i\vartheta \tag{2.48}$$

Here ϕ is the directed area of the circular sector OAB as follows.

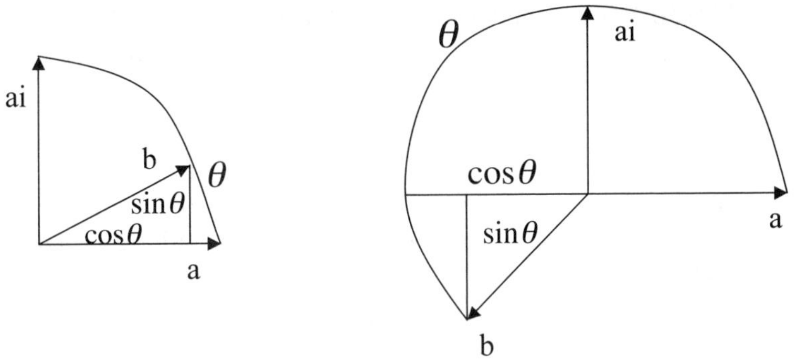

FIGURE 2.3
Linear and angular measure.

We know from the simple proportion that

$$\frac{\text{directed area of circular sector (with unit radius)}}{\text{arc length of the sector}}$$
$$= \frac{\text{directed area of unit circle}}{\text{circumference of circle}} = \frac{\pi i}{2\pi} = \frac{i}{2}.$$

So, we can write that, in the case under consideration, the directed area of circular sector OAB $= (i/2)\vartheta$.

Then the bivector ϕ defined in (2.48) can be written as

$$\Phi/2 = i\vartheta/2 = \text{directed area of the circular sector OAB.} \tag{2.49}$$

Here i specifies the plane of the angle $|\Phi/2|$ and specifies "areal magnitude" of the angle. The sign of ϑ in (2.49) is determined by the orientation assigned to the unit pseudoscalar i as shown in Figure 2.4.

Using Equation 2.49 we can write Equation 2.45 as

$$z = ab = \pm e^{\phi/2}, \tag{2.50}$$

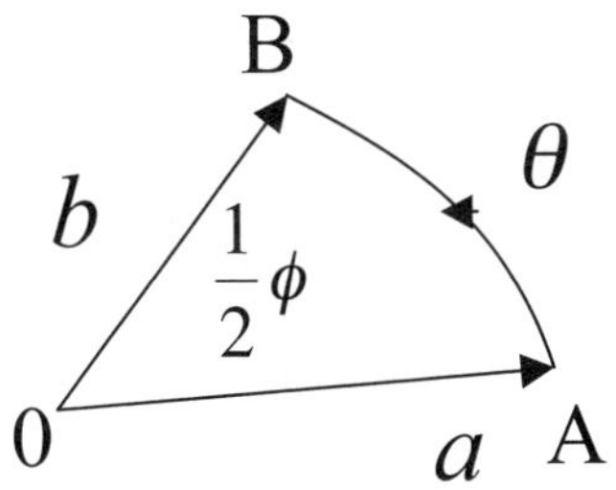

FIGURE 2.4
Angle and area of a circular sector.

where $\phi/2$ encodes the areal magnitude of the angle ϑ(in radian). This may be regarded as a functional relation of the bivector ϕ to the vectors $\hat{a}$ and $\hat{b}$.

We know that

$$z = ab = \pm e^{i\vartheta} \tag{2.45}$$

is a spinor of the i-plane. Each spinor z (with $|z| = 1$) determines a rotation in the i-plane. This is exemplified by the fact that the spinor z rotates each vector a in the i-plane into a vector b according to the equation

$$b = az = ae^{i\vartheta}. \tag{2.51}$$

Thus the exponential function $\exp(i\vartheta)$ represents a rotation in the i-plane as a function of the angle of rotation.

From the operational interpretation of $\exp(i\vartheta)$ one can obtain some important properties of the exponential function without any reference to its algebraic definition, which we discuss in the next section.

1. A rotation through a right angle ($\pi/2$) is represented by the unit bivector i, and hence

$$e^{i\pi/2} = i. \tag{2.52}$$

2. A rotation through two right angles π reverses the direction of a vector, and then

$$e^{i\pi} = -1. \tag{2.53}$$

3. A rotation through four right angles 2π gives the identity transformation of vectors represented by the multiplicative identity 1.
   Thus we may write

$$e^{2\pi i} = 1. \tag{2.54}$$

4. A rotation through an angle ϑ followed (or preceded) by a rotation through an angle φ is equivalent to a rotation through $(\vartheta + \varphi)$ and is expressed by

$$e^{i\vartheta}e^{i\varphi} = e^{i(\vartheta+\varphi)} = e^{i(\varphi+\vartheta)} = e^{i\varphi}e^{i\vartheta}. \tag{2.55}$$

5. Thus, if we consider n equal rotations in succession, each being through an angle ϑ, we get the well-known de Moivre's theorem:

$$\begin{aligned}(e^{i\vartheta})^n &= e^{i\vartheta}e^{i\vartheta}\ldots e^{i\vartheta} && (n \text{ factors})\\ &= e^{i\vartheta+i\vartheta+i\vartheta+\cdots+i\vartheta} && (n \text{ terms})\\ &= e^{\text{in}}\end{aligned} \tag{2.56}$$

From the foregoing discussion it is evident that the exponential function is a powerful means of describing rotations of vectors.

2.6 Exponential Functions of Multivectors

The exponential function of a multivector A of any arbitrary grade is denoted by $\exp(A)$ or e^A and is defined algebraically by the series expansion

$$\exp(A) = e^A = \sum_{k=0}^{\infty} A^k/k!$$

$$= 1 + A/1! + A^2/2! + \cdots + A^k/k! + \cdots, \tag{2.57}$$

if $|A|$ has a definite magnitude.

The series (2.57) can be shown to be absolutely convergent for all values of A provided $|A|$ has a definite magnitude. So, it may be extended to general multivectors.

The exponential function (2.57) is completely defined in terms of the basic operations of addition and multiplication (geometric product), which determine all the properties of the function. By using the closure property of the geometric algebra under the operations of addition and multiplication (geometric product) it can be shown that $\exp(A)$ is a definite multivector.

In particular, if the multivector A be an element of any algebra or subalgebra, such as $\mathcal{G}_3$, $\mathcal{G}_3^+$, $\mathcal{G}_2$, or $\mathcal{G}_2^+$, then the multivector $\exp(A)$ must be an element of the same algebra or subalgebra.

Now we prove a theorem demonstrating the additive rule for exponential function.

THEOREM 2.1
Prove the additive rule

$$e^A e^B = e^{A+B}, \tag{2.58}$$

if and only if $AB = BA$.

PROOF If $AB = BA$, then we have the identity

$$\sum_{m=0}^{\infty}(A^m/m!)\sum_{n=0}^{\infty}(B^n/n!) = \sum_{n=0}^{\infty}\sum_{k=0}^{n}(A^{n-k}B^k)/(n-k)!k! \tag{2.59}$$

By using the above identity, we can write

$$e^A e^B = \sum_{n=0}^{\infty}\sum_{k=0}^{n}(A^{n-k}B^k)/(n-k)!k! \tag{2.60}$$

By the binomial expansion we have

$$(A+B)^n = \sum_{k=0}^{n}[n!/(n-k)!k!](A^{n-k}B^k). \tag{2.61}$$

Thus, from (2.60) and (2.61) we can write

$$e^A e^B = \sum_{n=0}^{\infty}(A+B)^n/n! = e^{A+B} \tag{2.62}$$

The hyperbolic cosine and sine functions are defined, respectively, by the usual series expansion

$$\begin{aligned} coshA &= \sum_{k=0}^{\infty} A^{2k}/(2k)! \\ &= 1 + A^2/2! + A^4/4! + \cdots, \end{aligned} \tag{2.63}$$

$$\begin{aligned} sinhA &= \sum_{k=0}^{\infty} A^{2k+1}/(2k+1)! \\ &= A + A^3/3! + A^5/5! + \cdots, \end{aligned} \tag{2.64}$$

Adding (2.63) and (2.64) we obtain

$$\exp(A) = coshA + sinhA. \tag{2.65}$$

This shows that (2.63) and (2.64) are, respectively, "even" and "odd" parts of the exponential series. The multivector A is called the argument of each of the functions in (2.65).

The cosine and sine functions are defined, respectively, by the usual series expansions:

$$cosA = \sum_{k=0}^{\infty}(-1)^k A^{2k}/(2k)! = 1 - A^2/2! + A^4/4! - A^6/6! + \cdots, \tag{2.66}$$

$$sinA = \sum_{k=0}^{\infty}(-1)^k A^{2k+1}/(2k+1)! = A - A^3/3! + A^5/5! - A^7/7! + \cdots \tag{2.67}$$

If I is a multivector such that

$$I^2 = -1, \qquad IA = AI, \tag{2.68}$$

then, replacing A by IA in (2.63) and (2.64), we can express

$$coshIA = cosA, \tag{2.69}$$

$$sinhIA = IsinA, \tag{2.70}$$

$$e^{IA} = cosA + IsinA \tag{2.71}$$

Equation 2.71 is a generalization of (2.46). If A is a multivector of grade 0 (i.e., scalar) the definitions (2.66) and (2.67) for the trigonometric functions reduce, respectively, to the usual series expansions for cosine and sine of angles in radian measure. It is to be noted that the series expansions of exponential functions require that the angles be measured in radians.

Taking $i\vartheta$ for the multivector A in (2.57) and (2.71), we have from (2.45) the following series expansion of the geometric product of two unit vectors $\hat{a}$ and $\hat{b}$ in terms of their relative angle ϑ:

$$ab = e^{i\vartheta} = \sum_{k=0}^{\infty}(i\vartheta)^k/k! = 1 + i\vartheta - \vartheta^2/2! - i\vartheta^3/3! + \cdots, \tag{2.72}$$

References

1. D. Hestenes, *New Foundation for Classical Mechanics* (Reidel, Boston, 1986).

3

Euclidean Plane

3.1 The Algebra of Euclidean Plane

Following David Hestenes [1] we start with the vector equation for an oriented line. Every vector a determines a unique oriented line. This means that any vector x that is a scalar multiple of the vector a lies on the oriented line determined by a:

$$x = \alpha a, \tag{3.1}$$

where α is an arbitrary scalar. Equation 3.1 is said to be a parametric equation for the a-line.

A vector x is said to be *positively directed* or *negatively directed* relative to the vector a according as $x \cdot a > 0$ or < 0. This distinction that defines the positive and negative vectors is called the *orientation* or sense of the a-line.

The unit vector $\hat{a} = a\,|a|^{-1}$ is called the direction of the a-line, whereas $\hat{a}$ gives the opposite orientation for the line.

The parametric equation for the a-line can as well be written as

$$x = \beta \hat{a}, \tag{3.2}$$

where β is an arbitrary scalar.

The outer multiplication of (3.1) by vector a gives

$$x \wedge a = 0. \tag{3.3}$$

This is a nonparametric equation for the a-line. One can also write Equation 3.3 as

$$x \wedge \hat{a} = 0. \tag{3.4}$$

Now we can prove the following theorem.

THEOREM 3.1
Prove that the equation

$$x \wedge a = 0$$

has the solution set

$$x = \alpha a\,.$$

PROOF By definition of the geometric product we have

$$\begin{aligned} xa &= x \cdot a + x \wedge a \\ &= x \cdot a. \quad (\text{because } x \wedge a = 0) \end{aligned}$$

Multiplying the above equation on the right by

$$a^{-1} = aa^{-1}a^{-1} = aa^{-2},$$

we get

$$\begin{aligned} xaa^{-1} &= x \cdot aaa^{-2} \\ &= x \cdot aa^{-2}a \qquad (\text{because } aa^{-2} = a^{-2}a) \\ &= x \cdot a^{-1}a. \end{aligned}$$

Then

$$x = (x \cdot a^{-1})a$$

or

$$x = \alpha a$$

where we set

$$\alpha = x \cdot a^{-1}.$$

Hence the theorem.

1. Two-dimensional vector space
   In an analogous way we pass on to the algebraic description of a plane.
   For a nonzero bivector B, the set of all vectors x that satisfy the equation

   $$x \wedge B = 0 \tag{3.5}$$

   is said to be a two-dimensional vector space, and Equation 3.5 is referred to as representing the B-plane.
   The unit bivector i given by

   $$B = \mathcal{B}i, \tag{3.6}$$

   where $\mathcal{B}$ is a scalar and is called the direction of the B-plane. Then i determines an orientation of the B-plane and i gives the opposite orientation for the plane.
   Substituting (3.6) into (3.5), we get

   $$x \wedge i = 0. \tag{3.7}$$

This shows that every bivector that is a scalar (nonzero) multiple of i determines the same plane as i with same or opposite orientation according as the scalar is positive or negative.

2. Parametric equation for the i-plane
We express the equation for the i-plane

$$x \wedge i = 0 \tag{3.7}$$

into the parametric form.

First, we factorize i as the product of two unit orthogonal vectors σ_1 and σ_2:

$$i = \sigma_1\sigma_2 = \sigma_1 \wedge \sigma_2 = -\sigma_2 \wedge \sigma_1 = -\sigma_2\sigma_1, \tag{3.8}$$

where

$$\sigma_1 \cdot \sigma_2 = 0, \quad \sigma_1^2 = \sigma_2^2 = 1. \tag{3.9}$$

By the definition of the geometric product and using Equation 3.7, we obtain

$$\begin{aligned} xi &= x \cdot i \\ &= x \cdot (\sigma_1 \wedge \sigma_2) && \text{[by using (3.8)]} \\ &= (x \cdot \sigma_1)\sigma_2 - (x \cdot \sigma_2)\sigma_1 && \text{[by using (1.23)].} \end{aligned}$$

Multiplying the above equation on the right by the reverse of i, i.e., by $\tilde{i} = \sigma_2\sigma_1$, we get

$$\begin{aligned} xi\tilde{i} &= (x \cdot \sigma_1)\sigma_2\sigma_2\sigma_1 - (x \cdot \sigma_2)\sigma_1\sigma_2\sigma_1 \\ &= (x \cdot \sigma_1)\sigma_1 + (x \cdot \sigma_2)\sigma_2\sigma_1\sigma_1 && \text{[by using (3.8, 9)]} \end{aligned}$$

and being

$$i\tilde{i} = \sigma_1\sigma_2\sigma_2\sigma_1 = \sigma_1\sigma_1 = 1,$$

we have

$$x = (x \cdot \sigma_1)\sigma_1 + (x \cdot \sigma_2)\sigma_2. \tag{3.10}$$

By setting

$$x_1 = (x \cdot \sigma_1), \quad \text{and} \quad x_2 = (x \cdot \sigma_2), \tag{3.11}$$

Equation 3.10 can be written as

$$x = x_1\sigma_1 + x_2\sigma_2. \tag{3.12}$$

Equation 3.12 represents the parametric equation for the i-plane, where the scalars x_1 and x_2 are the rectangular components of the vector x with respect to

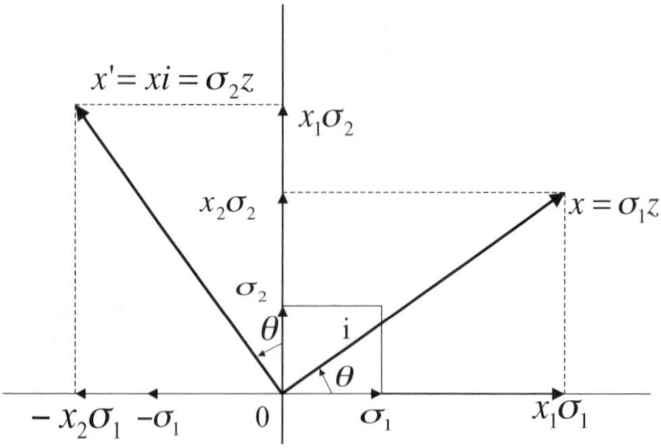

FIGURE 3.1
Diagram of the i-plane of vectors (real plane).

the basis $\{\sigma_1, \sigma_2\}$. Orthogonal vectors σ_1, σ_2 are represented by perpendicular line segments as shown in the Figure 3.1. A typical vector x is represented by a directed line segment, whereas the unit pseudoscalar i is represented by a plane segment of unit-directed area. So, one can sum up that i is the directed area of the plane segment and that the directed area of every plane segment in the i-plane is a scalar multiple of i.

3.2 Geometric Interpretation of a Bivector of Euclidean Plane

Now we shall show that the unit bivector i has two distinct geometric interpretations corresponding to two basic properties of the plane:

1. First, as stated earlier, it is the unit of a directed area representing the direction of the plane.
2. Second, it is the generator of rotations in the plane. The first interpretation is exemplified by Equation 3.8, which expresses that i is the product of two orthogonal unit vectors σ_1 and σ_2.

Next, in order to exhibit the second interpretation, we multiply Equation 3.8 on the left by σ_1 and σ_2, respectively, to obtain

$$\sigma_1 i = \sigma_1\sigma_1\sigma_2 = (\sigma_1\sigma_1)\sigma_2 = \sigma_2, \tag{3.13}$$

$$\sigma_2 i = \sigma_2\sigma_1\sigma_2 = -\sigma_1(\sigma_2\sigma_2) = -\sigma_1. \tag{3.14}$$

Equation (3.13) tells us that the multiplication of σ_1 on the right by the unit bivector i, also called the unit pseudoscalar, transforms σ_1 into σ_2. Because σ_1

and σ_2 are two unit orthogonal vectors, this transformation is a pure rotation of σ_1 through a right angle. Similarly, Equation 3.14 exemplifies a pure rotation of σ_2 through a right angle into $-\sigma_1$.

Substitution of (3.13) into (3.14) gives

$$(\sigma_1 i)i = -\sigma_1$$

or

$$\sigma_1 i^2 = -\sigma_1, \tag{3.15}$$

which states that two consecutive rotations through right angles reverse the direction of a vector. This explicity provides a geometric interpretation for the equation

$$i^2 = -1 \tag{3.16}$$

when i and -1 are both regarded as multiplicative operators on vectors.

The multiplication of any vector x in the i-plane on the right by i rotates x by a right angle into the vector x' given by

$$x' = xi = (x_1\sigma_1 + x_2\sigma_2)i = x_1\sigma_1 i + x_2\sigma_2 i$$

or

$$x' = x_1\sigma_2 - x_2\sigma_1 \quad \text{[by using (3.13) and (3.14)].} \tag{3.17}$$

The relation between x and x' is represented in the Figure 3.1.

We note in passing that the right multiplication by i rotates vectors counterclockwise by a right angle. A positive orientation of a plane corresponds to a counterclockwise rotation, whereas a negative orientation corresponds to a clockwise rotation.

3.3 Spinor *i*-Plane

The geometric interpretation for the equation

$$i^2 = -1 \tag{3.16}$$

gives rise to the construction of a spinor i-plane as given below.

We define a spinor of the i-plane to be a quantity obtained by the geometric product of two vectors in the i-plane.

Thus, from (3.8) and (3.9) we obtain a spinor z defined by the geometric product of σ_1 and x:

$$z = \sigma_1(x_1\sigma_1 + x_2\sigma_2) = x_1\sigma_1^2 + x_2\sigma_1\sigma_2. \tag{3.17}$$

Then,

$$z = x_1 + i x_2, \tag{3.18}$$

which is commonly called complex number.

It is to be noted that besides the algebraic property

$$i^2 = -1$$

ascribed to the traditional unit imaginary, our i is a bivector, also known as the unit pseudoscalar for the i-plane. So, it has geometric and algebraic properties beyond those traditionally accorded to imaginary numbers.

Analogously with the Argand diagram of the complex plane we construct the diagram of the spinor i-plane with two perpendicular axes, scalar and pseudoscalar.

As the unit pseudoscalar i is more than the unit imaginary $\sqrt{-1}$, so, also the spinor z represented by (3.18) is more than a complex number.

The set of all spinors of the form (3.18) is a two-dimensional plane called the spinor plane. The elements of a spinor plane can be represented by directed line segments or points in the diagram as shown in Figure 3.2.

Here we choose the vector σ_1 in the vector plane to construct a unique spinor z in the spinor plane (Figure 3.2) formed by the geometric product of σ_1 with each vector x given by (3.18). This, in fact, means that σ_1 distinguishes a line on the vector plane that is associated with the scalar axis in the spinor plane.

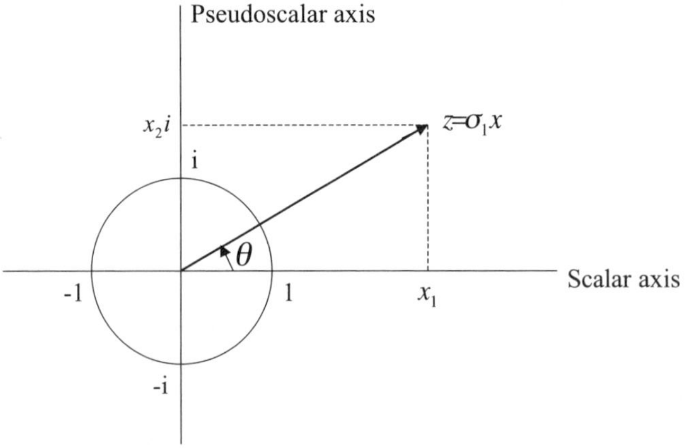

FIGURE 3.2
Diagram for the spinor i-plane. Each point in the spinor plane represents a rotation–dilation. Points on the unit circle represent pure rotations, whereas points on the positive scalar axis represent pure dilations.

3.3.1 Correspondence between the *i*-Plane of Vectors and the Spinor Plane

From the foregoing, one notices that each vector x in the i-plane of vectors determines a unique spinor z in the spinor plane as given by

$$z = \sigma_1 x = \sigma_1(x_1\sigma_1 + x_2\sigma_2) = x_1 + ix_2. \tag{3.19}$$

Conversely, each spinor z in the spinor plane determines a unique vector x in the vector plane as follows:

$$\begin{aligned} \sigma_1 x &= z, \\ \sigma_1^2 x &= \sigma_1 z \\ x &= \sigma_1 z. \end{aligned} \tag{3.20}$$

3.4 Distinction between Vector and Spinor Planes

The elements of the vector and spinor planes have different algebraic properties because of two distinct interpretations of i in them, which endow different geometric significance to each of the planes. The different interpretations of i that distinguish the two planes are stated below:

1. First, the interpretation of i as the unit directed area representing the direction of the vector plane is indicated in Figure 3.1 and Figure 3.2.
2. Second, the operator interpretation of i as a rotation of vectors through a right angle is indicated in Figure 3.2 by the right angle that the i-axis (pseudoscalar axis) makes with the scalar axis. This leads to an operator interpretation for all spinors as exemplified by Equation 3.20 (see Figure 3.2). So, the operator i may quite plausibly be called the generator of rotations.

Now we elaborate the interpretation of the equation

$$x = \sigma_1 z \tag{3.20}$$

and its representation as given in Figure 3.1 and Figure 3.2. Operating by right multiplication on the vector σ_1, the spinor z transforms σ_1 into a vector x, yielding a rotation of σ_1 through an angle

$$\vartheta = \tan^{-1}(x_2/x_1)$$

together with a dilation of σ_1 by an amount

$$|z| = \left(x_1^2 + x_2^2\right)^{1/2}.$$

As our choice of σ_1 is arbitrary, we can conclude that the spinor z must have the same effect on every vector on the i-plane. So, each spinor of the spinor plane can be regarded as an algebraic representation of a rotation–dilation.

Next, we discuss in detail the algebraic representation of spinors as rotation–dilation.

Let us consider a vector x and spinor z given by

$$x = \alpha_1\sigma_1 + \alpha_2\sigma_2, \tag{3.21}$$

$$z = \beta_1 + \beta_2 i, \tag{3.22}$$

where $\alpha_1, \beta_1, \alpha_2, \beta_2$ are positive scalars. By the right multiplication on the vector x by the spinor z, we get

$$\begin{aligned} xz &= (\alpha_1\sigma_1 + \alpha_2\sigma_2)(\beta_1 + \beta_2 i) \\ &= (\alpha_1\beta_1 - \alpha_2\beta_2)\sigma_1 + (\alpha_1\beta_2 + \alpha_2\beta_1)\sigma_2. \end{aligned} \tag{3.23}$$

$$\begin{aligned} |xz| &= [(\alpha_1\beta_1 - \alpha_2\beta_2)^2 + (\alpha_1\beta_2 +^{1/2} \alpha_2\beta_1)^2] \\ &= \left[\left(\alpha_1^2 + \alpha_2^2\right)\left(\beta_1^2 + \beta_2^2\right)\right]^{1/2} = |x||z|. \end{aligned} \tag{3.24}$$

The above equations show that the spinor z transforms x into a vector xz, yielding a rotation of x through an angle ϑ given by

$$\vartheta = \tan^{-1}[(\alpha_1\beta_2 + \alpha_2\beta_1)/(\alpha_1\beta_1 - \alpha_2\beta_2)] - \tan^{-1}(\alpha_2/\alpha_1) \tag{3.25}$$

together with a dilation of x by an amount

$$|z| = \left(\beta_1^2 + \beta_2^2\right)^{1/2}. \tag{3.26}$$

Case 1: For points on the scalar axis excluding the points $(1, 0)$ and $(-1, 0)$, we have $\alpha_2 = \beta_2 = 0$.

In this case

$$x = \pm\alpha_1\sigma_1, \quad z = \pm\beta_1, \quad xz = \alpha_1\beta_1\sigma_1. \tag{3.27}$$

Equation 3.27 show that whereas the points on the positive scalar axis represent pure dilation by an amount $|z| = \beta_1$, those on the negative scalar axis represent rotation through an angle π together with dilation by an amount $|z| = \beta_1$.

Case 2: For points on the pseudoscalar axis excluding the points $(0, i)$ and $(0, -i)$, we have $\alpha_1 = \beta_1 = 0$. In this case

$$x = \pm\alpha_2\sigma_2, \quad z = \pm\beta_2 i, \quad xz = -\alpha_2\beta_2\sigma_1. \tag{3.28}$$

Equation 3.28 shows that whereas the points on the positive pseudoscalar axis represent rotation through an angle $\pi/2$ together with dilation by an amount $|z| = \beta_2$, those on the negative pseudoscalar axis represent rotation through an angle $3\pi/2$ together with dilation by an amount $|z| = \beta_2$.

Case 3: For points on the unit circle excluding the points (1,0), (−1,0), and (0 − i), we have

$$\alpha_1^2 + \alpha_2^2 = 1 = \beta_1^2 + \beta_2^2 \quad \text{and} \quad |xz| = 1 = |x|. \tag{3.29}$$

This shows that spinors in this case yield a rotation of x through an angle ϑ given by

$$\vartheta = \tan^{-1}[(\alpha_1\beta_2 + \alpha_2\beta_1)/(\alpha_1\beta_1 - \alpha_2\beta_2)] - \tan^{-1}(\alpha_2/\alpha_1), \tag{3.30}$$

but no dilation of x, i.e., the corresponding points on the unit circle represent pure rotations.

Case 4: For points (1,0) and (−1,0) where the scalar axis meets the unit circle, we have

$$x = \pm\sigma_1, \quad z = \pm 1, \quad xz = \sigma_1. \tag{3.31}$$

This tells us that the point (1,0) represents no rotation–dilation, whereas the point (−1,0) represents a pure rotation through an angle π.

Case 5: For the points (0, i) and (0,−i) where the pseudoscalar axis meets the unit circle, we have

$$x = \pm\sigma_2, \quad z = \pm i, \quad xz = -\sigma_1. \tag{3.32}$$

This tells us that the points (0, i) and (0, $-i$) represent pure rotations through angles $\pi/2$ and $3\pi/2$, respectively.

3.4.1 Some Observations

1. It is important to note that the reversion of a spinor corresponds to complex conjugation as shown below:

$$\begin{aligned} \tilde{z} &= (\sigma_1 x)^{\sim} = x\sigma_1 = (x_1\sigma_1 + x_2\sigma_2)\sigma_1 \\ &= x_1\sigma_1^2 + x_2\sigma_2\sigma_1 = x_1\sigma_1^2 - x_2\sigma_1\sigma_2 \\ \tilde{z} &= x_1 - x_2 i. \end{aligned} \tag{3.33}$$

2. The notation representing the modulus of the spinor z agrees with the conventional notation for the modulus of a complex number, i.e.,

$$|z|^2 = z\tilde{z} = \tilde{z}z = x_1^2 + x_2^2. \tag{3.34}$$

3. The separation of a spinor into scalar and pseudoscalar parts can be done analogously to separating a complex number into real and imaginary parts:

$$x_1 = Re\{z\} = (z + \tilde{z})/2, \tag{3.35a}$$

$$x_2 = Im\{z\} = (z - \tilde{z})/2i. \tag{3.35b}$$

3.5 The Geometric Algebra of a Plane

In the foregoing text we have the concept of vector space and spinor space with different geometric significances. Both the spaces are linear spaces; the vector space is a linear space of vectors, and the spinor space is a linear space of spinors. Unlike conventional mathematics, the term vector space is not synonymous with linear space in geometric algebra. This can be traced in the following arguments. Geometric algebra ascribes some special properties to vectors by introducing (1) the geometric product of vectors with associative property and (2) the outer product of vectors defining a pseudoscalar for the plane, regarded as a generator of rotations in the plane. This, in fact, restricts the use of vectors in geometric algebra to some precise sense. As geometric algebra consists of two kinds of quantities, i.e., vectors, which are odd multivectors, and spinors, which are even multivectors, the linear space is constructed out of vectors and spinors. Thus, the linear space is the sum of two linear spaces, vector and spinor. So, we restrict the term vector space to a linear space of vectors in geometric algebra. The total linear space is not a vector space.

Now, we develop the geometric algebra of a plane. First, we introduce the concept of linear space in its usual general sense.

A set of quantities $A_1, A_2, \ldots, A_n$ is said to be linearly dependent if there exist n scalars $\alpha_1, \alpha_2, \ldots, \alpha_n$, not all zero, such that

$$\alpha_1 A_1 + \alpha_2 A_2 + \cdots + \alpha_n A_n = 0. \tag{3.36}$$

Otherwise the set is linearly independent.

If the $A_k (k = 1, 2, \ldots, n)$ are linearly independent, then the set of all linear combinations of the A_k is said to constitute an n-dimensional linear space, and the $\{A_k\}$ is said to be a basis for the linear space.

Consider the equation for the i-plane

$$x = x_1\sigma_1 + x_2\sigma_2, \tag{3.12}$$

where the orthonormal vectors σ_1 and σ_2 form a basis for the plane. In geometric algebra, the geometric product of the vectors σ_1 and σ_2 generates two more basis elements, namely, the unit scalar $1 = \sigma_1^2 = \sigma_2^2$ and the unit pseudoscalar $i = \sigma_1\sigma_2$. Thus, the geometric algebra of the i-plane is generated by two orhtonormal vectors σ_1 and σ_2 and is spanned by four linearly independent quantities $\{1, \sigma_1, \sigma_2, i\}$. So, every multivector A can be expressed as a linear combination

$$A = \alpha_0 1 + \alpha_1\sigma_1 + \alpha_2\sigma_2 + \alpha_3 i \tag{3.37}$$

with scalar coefficients $\alpha_0, \alpha_1, \alpha_2$, and α_3.

The set of all multivectors generated from the vectors of the i-plane by the addition of vectors and by the addition of the geometric product of vectors is the geometric algebra of the i-plane and it is denoted by $\mathcal{G}_2(i)$ or simply by $\mathcal{G}_2$.

The subscript "2" refers both to the dimension of the plane and the grade of the pseudoscalar. As the four unit multivectors $\{1, \sigma_1, \sigma_2, i\}$ make up a basis for this space, the algebra $\mathcal{G}_2$ is a four-dimensional linear space.

It is obvious from (3.37) that any multivector A in $\mathcal{G}_2$ can be expressed as the sum of a vector $a = \alpha_1\sigma_1 + \alpha_2\sigma_2$ and a spinor $z = \alpha_0 + \alpha_3 i$, i.e.,

$$A = a + z. \tag{3.38}$$

The vectors are odd multivectors, and spinors are even multivectors. So, $\mathcal{G}_2$ can be expressed as the sum of two linear spaces

$$\mathcal{G}_2 = \mathcal{G}_2^- + \mathcal{G}_2^+, \tag{3.39}$$

where $\mathcal{G}_2^-$ is the two-dimensional linear space of vectors, and $\mathcal{G}_2^+$ is the two-dimensional linear space of spinors, both being closed under multiplication. Of the two linear spaces $\mathcal{G}_2^-$ and $\mathcal{G}_2^+$, only $\mathcal{G}_2^-$ is a vector space. $\mathcal{G}_2$ is a four-dimensional linear space but not a vector space.

The two-dimensional vector space $\mathcal{G}_2^-$ is also the i-plane itself. As the scalar product is defined in it, this vector space is Euclidean and is denoted by E_2.

The dimension of a linear space is the number of linearly independent elements of the space. However, for a vector space and its algebra, its dimension is synonymous with the grade of its pseudoscalar. A vector space of dimension "2" is determined by a pseudoscalar of grade "2" and *vice versa*.

We note in passing that the algebra of complex numbers appears with a geometric interpretation as the subalgebra $\mathcal{G}_2^+$ of even multivectors in $\mathcal{G}_2$.

References

1. D. Hestenes, *New Foundations for Classical Mechanics* (Reidel, Boston, 1986).

4

The Pseudoscalar and Imaginary Unit

4.1 The Geometric Algebra of Euclidean 3-Space

Following D.Hestenes [1] we develop and analyze the algebra of Euclidean 3-space. This can be done in the same way as in the analysis of the algebra of E_2 in the previous chapters.

Let i be a unit trivector in E_3, preferably called a unit pseudoscalar for E_3. The set of all vectors x that satisfies the equation

$$x \wedge i = 0 \tag{4.1}$$

is the Euclidean three-dimensional vector space E_3. Scalar multiples of i are called pseudoscalars for E_3.

One can factorize i as a product of three orthonormal vectors σ_1, σ_2, and σ_3:

$$i = \sigma_1\sigma_2\sigma_3 = \sigma_1 \wedge \sigma_2 \wedge \sigma_3, \tag{4.2}$$

where

$$\sigma_1^2 = \sigma_2^2 = \sigma_3^2 = 1, \tag{4.3}$$

$$\sigma_i \cdot \sigma_j = 0 \quad \text{if} \quad i \neq j (i, j = 1, 2, 3). \tag{4.4}$$

We assume that the orthonormal vectors σ_1, σ_2, and σ_3 form a right-handed or dextral set of vectors forming a basis for E_3. Equation 4.2 specifies a definite relation of the pseudoscalar i to the dextral set of vectors. This signifies that i is the dextral or right-handed unit pseudoscalar. By reversing the directions of the vector σ_k we get a left-handed set of vectors $\{-\sigma_k\}$ and the left-handed unit pseudoscalar $(-\sigma_1)(-\sigma_2)(-\sigma_3) = -i$.

From Equation 4.3 and Equation 4.4 we can also write

$$\sigma_i\sigma_j = \sigma_i \wedge \sigma_j = -\sigma_j \wedge \sigma_i = -\sigma_j\sigma_i \quad \text{if} \quad i \neq j. \tag{4.5}$$

We deduce the parametric equation for E_3 from Equation 4.1. By the definition of the geometric product we have from (4.1),

$$xi = x \cdot i.$$

By using (4.2), this can be written as

$$x\sigma_1\sigma_2\sigma_3 = x \cdot (\sigma_1 \wedge \sigma_2 \wedge \sigma_3).$$

Then, by applying the reduction formula (1.60) to the right-hand side of the above equation, we obtain

$$xi = x\sigma_1\sigma_2\sigma_3 = x \cdot \sigma_1\sigma_2 \wedge \sigma_3 - x \cdot \sigma_2\sigma_1 \wedge \sigma_3 + x \cdot \sigma_3\sigma_1 \wedge_2 \sigma.$$

Multiplying the above on the right by $\tilde{i} = \sigma_3\sigma_2\sigma_1$, we get, by using (4.3) and (4.5)

$$\begin{aligned}
xi\tilde{i} &= x \cdot \sigma_1\sigma_2\sigma_3\sigma_3\sigma_2\sigma_1 - x \cdot \sigma_2\sigma_1\sigma_3\sigma_3\sigma_2\sigma_1 + x \cdot \sigma_3\sigma_1\sigma_2\sigma_3\sigma_2\sigma_1 \\
&= x \cdot \sigma_1\sigma_2\sigma_2\sigma_1 - x \cdot \sigma_2\sigma_1\sigma_2\sigma_1 - x \cdot \sigma_3\sigma_1\sigma_2\sigma_2\sigma_3\sigma_1 \\
&= x \cdot \sigma_1\sigma_1 + x \cdot \sigma_2\sigma_1\sigma_1\sigma_2 - x \cdot \sigma_3\sigma_1\sigma_3\sigma_1 \\
&= x \cdot \sigma_1\sigma_1 + x \cdot \sigma_2\sigma_2 + x \cdot \sigma_3\sigma_1\sigma_1\sigma_3 \\
&= x \cdot \sigma_1\sigma_1 + x \cdot \sigma_2\sigma_2 + x \cdot \sigma_3\sigma_3.
\end{aligned}$$

Thus, we have

$$x = (x \cdot \sigma_1)\sigma_1 + (x \cdot \sigma_2)\sigma_2 + (x \cdot \sigma_3)\sigma_3,$$

which may be expressed as

$$x = x_1\sigma_1 + x_2\sigma_2 + x_3\sigma_3, \tag{4.6}$$

where $x_k = x \cdot \sigma_k, k = 1, 2, 3$. The scalars x_k are the rectangular components of the vector x with respect to the basis $\{\sigma_1, \sigma_2, \sigma_3\}$. Equation 4.6 represents the parametric equation for E_3.

Orthogonal vectors σ_1, σ_2 and σ_3 are represented by perpendicular line segments as shown in Figure 4.1. The right-handed unit pseudoscalar i is represented in the figure by the oriented unit cube.

Multiplication of (4.2) by σ_1, σ_2 and σ_3 in succession on the left and also on the right gives three linearly independent bivectors, which we denote by i_1, i_2 and i_3:

$$i_1 = \sigma_1 i = \sigma_1\sigma_1\sigma_2\sigma_3 = \sigma_2\sigma_3 = i\sigma_1. \tag{4.7a}$$

$$i_2 = \sigma_2 i = \sigma_2\sigma_1\sigma_2\sigma_3 = -\sigma_1\sigma_2\sigma_2\sigma_3 = -\sigma_1\sigma_3 = \sigma_3\sigma_1 = i\sigma_2. \tag{4.7b}$$

$$i_3 = \sigma_3 i = \sigma_3\sigma_1\sigma_2\sigma_3 = -\sigma_1\sigma_3\sigma_2\sigma_3 = \sigma_1\sigma_2\sigma_3\sigma_3 = \sigma_1\sigma_2 = i\sigma_3. \tag{4.7c}$$

Equations 4.7a, b, c imply that the unit pseudoscalar i commutes with all vectors of E_3. The unit bivectors i_1, i_2, and i_3 are represented in the Figure 4.1 by the oriented-plane segments.

The set of all multivectors generated from the vectors of E_3 by the addition of vectors and by the addition of the geometric products of vectors is the geometric algebra of E_3, and it is denoted by $\mathcal{G}_3(i)$ or simply by $\mathcal{G}_3$. The subscript '3' refers both to the dimension of the space and the grade of the pseudoscalar of E_3.

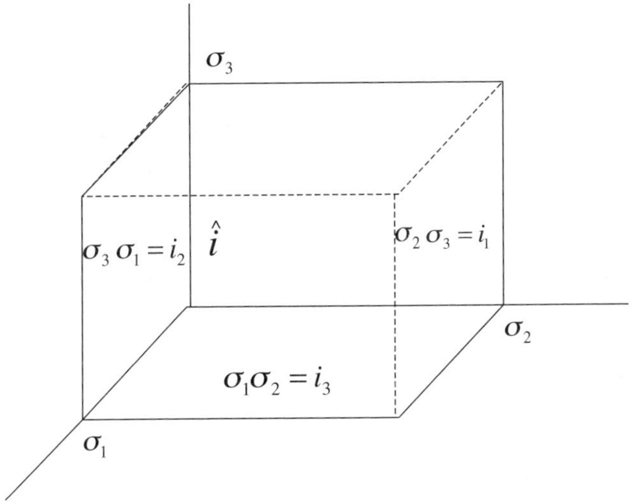

FIGURE 4.1
Orthonormal basis for E_3. Directed line segments represent unit vectors. Directed plane segments represent unit bivectors.The oriented cube represents the right-handed unit pseudoscalar i.

In the foregoing we have seen that in $\mathcal{G}_3$ the geometric product of the vectors σ_1, σ_2 and σ_3 generates five more basis elements, namely, the unit scalar $1 = \sigma_1^2 = \sigma_2^2 = \sigma_3^2$, the unit pseudoscalar $i = \sigma_1\sigma_2\sigma_3$, and three linearly independent unit bivectors $\sigma_2\sigma_3$, $\sigma_3\sigma_1$ and $\sigma_1\sigma_2$. So, the geometric algebra of E_3 is generated by the vectors σ_1, σ_2, and σ_3 and is spanned by eight linearly independent quantities

$$\begin{array}{cccc} 1, & \{\sigma_1, \sigma_2, \sigma_3\}, & \{\sigma_2\sigma_3, \sigma_3\sigma_1, \sigma_1\sigma_2\}, & i. \\ (\textit{scalar} — 1) & (\textit{vectors} — 3) & (\textit{bivectors} — 3) & (\textit{pseudoscalar} — 1) \end{array} \tag{4.8}$$

The algebra $\mathcal{G}_3$ is a linear space of eight dimensions. As $i\sigma_k$ $(k = 1, 2, 3)$ are the only linearly independent bivectors that can be obtained from σ_k by geometric product, any bivector B in $\mathcal{G}_3$ can be expressed as a linear combination:

$$\begin{aligned} B &= \mathcal{B}_1 i\sigma_1 + \mathcal{B}_2 i\sigma_2 + \mathcal{B}_3 i_3\sigma \\ &= i(\mathcal{B}_1\sigma_1 + \mathcal{B}_2\sigma_2 + \mathcal{B}_3\sigma_3), \end{aligned} \tag{4.9}$$

with scalar coefficients $\mathcal{B}_k$. Thus, the set of all bivectors in $\mathcal{G}_3$ is a three-dimensional linear space with a basis $\{i\sigma_1, i\sigma_2, i\sigma_3\}$.

From Equation 4.9 we notice that every bivector B in $\mathcal{G}_3$ is uniquely related to a vector

$$b = \mathcal{B}_1\sigma_1 + \mathcal{B}_2\sigma_2 + \mathcal{B}_3\sigma_3 \tag{4.10}$$

by the equation

$$B = ib \tag{4.11}$$

This relation is expressed by saying that the bivector B is the dual of the vector b.

In general we define the dual of any multivector M in $\mathcal{G}_3$ as the product iM, where i is the dextral unit pseudoscalar of E_3.

In view of the Axiom A.15 of Chapter 1, it follows from Equation 4.1 that $\mathcal{G}_3$ contains no nonzero k-vectors for $k \geq 4$. So, by definition (1.28), every multivector M in $\mathcal{G}_3$ can be written in the expanded form

$$M = M_0 + M_1 + M_2 + M_3, \tag{4.12}$$

where M_g ($g = 0, 1, 2, 3$) denotes a multivector of grade g. Thus, in this case we have

$$\begin{aligned} M_0 &= \alpha, \text{ the scalar part of } M, \\ M_1 &= a, \text{ the vector part of } M, \\ M_2 &= ib, \text{ the bivector part of } M \text{ expressed} \\ &\quad \text{as a dual of a vector } b, \text{ and} \\ M_3 &= i\beta, \text{ the pseudoscalar part of } M. \end{aligned}$$

Then, (4.12) can be put in the form

$$M = \alpha + a + ib + i\beta. \tag{4.13}$$

The eight unit multivectors $\{1, \sigma_k, i\sigma_k, i\}$ with $k = 1, 2, 3$ make up a basis for the algebra $\mathcal{G}_3$ defined by (4.8).

Denoting by $\mathcal{G}_3^k$ the subspace of k-vectors in $\mathcal{G}_3$, we may write:

$$\begin{aligned} \mathcal{G}_3^1 &= a \text{ three-dimensional space of vectors,} \\ \mathcal{G}_3^2 &= a \text{ three-dimensional space of bivectors,} \\ \mathcal{G}_3^3 &= a \text{ one-dimensional space of trivectors.} \end{aligned}$$

4.1.1 The Pseudoscalar of E_3

The pseudoscalar i that plays an important role in $\mathcal{G}_3$ has the following basic properties:

$$i^2 = -1 \tag{4.14a}$$

$$\widetilde{i} = -i \tag{4.14b}$$

$$iM = Mi \text{ for every multivector } M \text{ in } \mathcal{G}_3 \tag{4.14c}$$

$$a \wedge b \wedge c = \lambda i \tag{4.14d}$$

for any set of linearly independent vectors a, b, and c in $\mathcal{G}_3$ and any scalar λ. One must note that the scalar λ is positive if and only if the vectors make up a dextral set in the order given. Here the reversion of i corresponds to complex conjugation.

4.2 Complex Conjugation

As the properties (4.14) of the unit pseudoscalar are similar to those usually attributed to the unit imaginary in mathematics, the unit pseudoscalar in $\mathcal{G}_3$ is denoted by the symbol i. Nevertheless, there is not just one root of minus one in geometric algebra but many. The elements in a basis of the algebra endowed with geometric interpretations indicate that in $\mathcal{G}_3$ there are two distinct kinds of solutions of the equation $x^2 = -1$:

1. x is a pseudoscalar, whence $x = \pm i$, or
2. x is a bivector, whence $x = i_1$ or i_2 or i_3.

Appendix A: Some Important Results

1. Vector identities
   (a) $(a \wedge b) \cdot (c \wedge d) = b \cdot ca \cdot d - b \cdot da \cdot c = b \cdot (c \wedge d) \cdot a.$
   (b) $a \cdot (\sigma_1 \wedge \sigma_2 \wedge \sigma_3) = a \cdot \sigma_1 \sigma_2 \wedge \sigma_3 - a \cdot \sigma_2 \sigma_1 \wedge \sigma_3 + a \cdot \sigma_3 \sigma_1 \wedge \sigma_2.$
2. $a \wedge b \wedge c = 0$ if and only if $a, b, c,$ are linearly dependent.
3. Jacobi identity for vectors:
$$a \cdot (b \wedge c) + b \cdot (c \wedge a) + c \cdot (a \wedge b) = 0$$
4. If A_r is a multivector of grade r, then
$$A_r^{-1} = \tilde{A}/|A_r|^2.$$
5. If A, B are multivectors, then
$$\begin{aligned} &(a) \quad (AB)_0 = (BA)_0. \\ &(b) \quad (AB)_+ = (A)_+(B)_+ + (A)_-(B)_- \\ &(c) \quad (AB)_- = (A)_+(B)_- + (A)_-(B)_+ \\ &(d) \quad (\tilde{A})_r = (-1)^{(r/2)(r-1)}(A)_r. \end{aligned}$$
6. If $A = \alpha + i\beta + a + ib$, where α, β are scalars and a, b are vectors, then
$$\tilde{A} = \alpha - i\beta + a - ib$$
$$|\tilde{A}|^2 = \alpha^2 + \beta^2 + a^2 + b^2 = |A|^2.$$

7. If a and b are vectors in the plane of $i = \sigma_1\sigma_2$, then

$$a \wedge b = i \begin{vmatrix} a \cdot \sigma_1 & b \cdot \sigma_1 \\ a \cdot \sigma_2 & b \cdot \sigma_2 \end{vmatrix}$$

8. The expansion of a vector b in terms of its components $b_k = b \cdot \sigma_k$ is given by

$$b = b_k\sigma_k = b_1\sigma_1 + b_2\sigma_2 + b_3\sigma_3.$$

By this convention, the expansion of a bivector B in terms of its components

$$B_{ij} = \sigma_i \cdot B \cdot \sigma_j = (\sigma_j \wedge \sigma_i) \cdot B = -\mathfrak{B}_{ji}$$

can be written as

$$B = (1/2)\mathfrak{B}_{ij}\sigma_j \wedge \sigma_i = \mathfrak{B}_{12}\sigma_2 \wedge \sigma_1 + \mathfrak{B}_{23}\sigma_3 \wedge \sigma_2 + \mathfrak{B}_{31}\sigma_1 \wedge \sigma_3.$$

9. The product rule for Pauli matrices $\sigma_k(k = 1, 2, 3,)$ is

$$\sigma_k\sigma_l = \delta_{kl} + i\epsilon_{\text{klm}}\sigma_m.$$

Likewise, the product (geometric) rule for bivectors in Pauli algebra (see (4.7a, 7b, 7c)) i_k (k = 1, 2, 3,) is

$$i_k i_l = \delta_{kl} + \varepsilon_{klm} i_m.$$

References

1. D. Hestenes, *New Foundations for Classical Mechanics* (Reidel, Boston, 1986).

5

Real Dirac Algebra

5.1 Geometric Significance of the Dirac Matrices γ_μ

Before developing the geometric algebra of space-time or space-time algebra (STA) it would be in order for us to discuss the algebra of Dirac matrices γ_μ ($\mu = 0, 1, 2, 3$), endowing the geometric significance of the matrices γ_μ. This is so because, in Section 9.1 of Chapter 9 of this book, the Dirac equation is reformulated in STA without any reference to the complex number where the four Dirac γ_μ are considered to be linearly independent vectors belonging to an associative noncommutative division algebra.

The four gamma matrices of the Dirac theory over the complex numbers generate the complete algebra of 4×4 matrices. The 4×4 matrix algebra with the geometric interpretation induced by the conditions

$$(1/2)(\gamma_\mu \gamma_\nu + \gamma_\nu \gamma_\mu) = g_{\mu\nu} I, \tag{5.1a}$$

$$(1/4)Tr(\gamma_\mu \gamma_\nu) = g_{\mu\nu}, \tag{5.1b}$$

where the $g_{\mu\nu}(\mu, \nu = 0, 1, 2, 3)$ are the components of space-time metric tensor and I is the four-dimensional unit matrix, is called the algebra of Dirac matrices. Condition (5.1b) implies that the matrices γ_μ are irreducible. Conditions (5.1a, b) show that the matrices are traceless:

$$(1/4)Tr\gamma_\mu = 0. \tag{5.2}$$

In the conventional approach, the full geometric significance of the γ_μ can be understood only with the specification of their relation to the Dirac spinor. Essentially, the relation (5.1a) is independent on the assumption that the γ_μ are matrices. This can be realized by considering that the γ_μ belong to an associative noncommutative division algebra. This has been achieved by David Hestenes [1–5] by interpreting the γ_μ as vectors of a space-time frame with γ_0 the reference frame's 4-velocity, instead of as matrices. So the conditions (5.1a, b) for matrices correspond to the single equation

$$(1/2)(\gamma_\mu \gamma_\nu + \gamma_\nu \gamma_\mu) = \gamma_\mu \cdot \gamma_\nu = g_{\mu\nu} \tag{5.3}$$

for vectors.

The vectors γ_μ generate an associative algebra over the reals, which provides a direct and complete algebraic characterization of the geometric properties of Minkowski space-time. This associative algebra has been dubbed by Hestenes [1] as 'space-time algebra'.

As the geometric interpretation of the γ_μ of space-time algebra is independent of the notion of spinor, the γ_μ assumes a central position in the mathematical description of all physical systems in space-time, including the relativistic quantum theory.

5.2 Geometric Algebra of Space-Time

Earlier we have stated that the geometric algebra combines the algebraic structure of Clifford algebra with the explicit geometric meaning of its mathematical elements at its foundation. So, formally, it is Clifford algebra endowed with geometrical information of and physical interpretation to all mathematical elements of the algebra.

With the usual choice (+ - - -) for the metric of space-time, we consider four orthonormal vectors $\{\gamma_0, \gamma_1, \gamma_2, \gamma_3\}$ where

$$\gamma_0^2 = -\gamma_k^2 = 1, \quad (k = 1, 2, 3). \tag{5.3a}$$

The Clifford algebra of 'real' four-dimensional space-time is generated by four orthonormal vectors $\{\gamma_\mu\}$ and spanned by

$1,$	$\{\gamma_\mu\},$	$\{\sigma_k,\ i\sigma_k\},$	$\{i\gamma_\mu\},$	$i,$
1 *scalar*	4 *vectors*	6 *bivectors*	4 *trivectors*	1 *quadrivector*
grade 0	*grade* 1	*grade* 2	(or pseudovectors)	(or pseudoscalar)
			grade 3	*grade* 4

(5.4)

$$(\mu = 0, 1, 2, 3;\ k = 1, 2, 3),$$

where i is the unit pseudoscalar for space-time:

$$i = \gamma_0\gamma_1\gamma_2\gamma_3 = \sigma_1\sigma_2\sigma_3, \tag{5.5a}$$

$$\sigma_k = \gamma_k\gamma_0, \tag{5.5b}$$

$$i^2 = -1. \tag{5.5c}$$

In accordance with the notation adopted in current literature, we have taken i as the unit pseudoscalar for space-time. It will be shown later that, because of the relation (5.5a, b), i can also be taken as the unit pseudoscalar for three-dimensional Euclidean space. There could be no confusion for i representing two different geometric entities because the geometric meaning of i would be explicit from the context, whether it belongs to space-time or three-dimensional Euclidean space.

The algebra (5.4) is the space-time algebra (i.e., geometric algebra for space-time) or 'real Dirac algebra' having 16 components, and so is a 16-dimensional linear space. Real Dirac algebra is isomorphic to the algebra of Dirac matrices over the 'real' numbers.

In Dirac algebra,

$$i\gamma_\mu = -\gamma_\mu i. \tag{5.6}$$

A multivector M in Dirac algebra can be written (putting in evidence the parts with different grades) as

$$M = \alpha + a + B + ib + i\beta, \tag{5.7}$$

where α and β are scalars, a and b are vectors, and B is a bivector:

$$B = (1/2)B^{\mu\nu}\gamma_\mu \wedge \gamma_\nu. \tag{5.8}$$

In order to facilitate the decomposition of a multivector with respect to the basis $\{\gamma_\mu\}$, it is convenient to introduce a reciprocal basis $\{\gamma^\mu\}$, defined, as usual, by the relation

$$\gamma^\mu \cdot \gamma_\nu = \delta^\mu_\nu. \tag{5.9}$$

From the foregoing, one can have the unique relation

$$g_{\mu\nu} = (1/2)(\gamma_\mu\gamma_\nu + \gamma_\nu\gamma_\mu) = \gamma_\mu \cdot \gamma_\nu, \tag{5.10}$$

where $g_{\mu\nu}$ is the Minkowski metric (See Reference [1]).

The contravariant components of a vector $a = a^\nu\gamma_\nu$ are given, in view of (5.9), by

$$a^\mu = a \cdot \gamma^\mu. \tag{5.11a}$$

Likewise, the covariant components of the vector a are given by

$$a_\mu = a \cdot \gamma_\mu = g_{\mu\nu}a^\nu. \tag{5.11b}$$

Analogously, the components $B^{\mu\nu} = -B^{\nu\mu}$ of the bivector B in (5.8) are given by (see Equation 1.74)

$$B^{\mu\nu} = (\gamma^\mu \wedge \gamma^\nu) \cdot B = \gamma^\mu \cdot (\gamma^\nu \cdot B). \tag{5.12}$$

One may note in passing that the pseudovectors $i\gamma_\mu$ of Dirac algebra, which are dual of the vectors γ_μ, are trivectors as, for example, one can see that

$$i\gamma_3 = \gamma_0\gamma_1\gamma_2\gamma_3\gamma_3 = -\gamma_0\gamma_1\gamma_2. \tag{5.13}$$

In fact, the unit pseudoscalar i has the role of a multiplicative operator that determines the dual element of a proper multivector.

Multivector M in (5.7) of Dirac algebra can be decomposed as the sum of an even part M_+[grade 0 (scalar), grade 2 (bivector), and grade 4 (pseudoscalar)] and an odd part M_-[grade 1 (vector) and grade 3 (trivector)] as

$$M = M_+ + M_- \tag{5.14}$$

where

$$M_+ = \underset{(scalar)}{\alpha} + \underset{(bivector)}{B} + \underset{(pseudoscalar)}{i\beta} \tag{5.15}$$

$$M_- = \underset{(vector)}{a} + \underset{(pseudovector).}{ib} \tag{5.16}$$

The even multivectors or spinors (i.e., the sum of Clifford objects of even grade) form a subalgebra (of 8-dimensional linear space) of Dirac algebra, which is isomorphic to Pauli algebra. This is evident from the fact that even the Clifford objects of the basis (5.4) of Dirac algebra

$$1, \quad \{\sigma_k, i\sigma_k\}, \quad i \tag{5.17}$$

coincide with the basis of Pauli algebra because of the relation (5.5b), which in turn satisfies

$$\sigma_k\sigma_j = (1/2)(\sigma_k\sigma_j + \sigma_j\sigma_k) = -(1/2)(\gamma_k\gamma_j + \gamma_j\gamma_k) = \delta_{kj}, \tag{5.18}$$

the requirement of a basis in three-dimensional Euclidean space [6–10]. It is important to note that in Pauli algebra, because of the relation (5.5a)i becomes the unit pseudoscalar for three-dimensional Euclidean space and that i commutes with σ_k:

$$i\sigma_k = \sigma_k i, \tag{5.19}$$

indicating its similarity in character with the unit imaginary.

In geometric algebra, the three Pauli σ_k are no longer viewed as three matrix-valued components

$$\hat{\sigma}_1 = \begin{pmatrix} 0 & 1 \\ 1 & 0 \end{pmatrix}, \quad \hat{\sigma}_2 = \begin{pmatrix} 0 & -j \\ j & 0 \end{pmatrix}, \quad \hat{\sigma}_3 = \begin{pmatrix} 1 & 0 \\ 0 & -1 \end{pmatrix},$$

$$(j = \sqrt{-1}, \quad \text{the unit imaginary})$$

of a single isospace vector, but as three orthonormal basis vectors for three-dimensional Euclidean space (see Reference [3]). Likewise, the $\{\gamma_k\}$, $\{\sigma_k\}$ are to be interpreted geometrically as spatial vectors (space-time bivectors) and not as operators in an abstract spin space. 'Real' Pauli algebra is isomorphic to the algebra of Pauli matrices.

A new and important geometric significance, and interesting at that, can be assigned to the foregoing [8,9,11–14]. In space-time algebra, the four Dirac

γ_μ are no longer viewed as four matrix-valued components

$$\hat{\gamma}_0 = \begin{pmatrix} \mathrm{I} & 0 \\ 0 & -\mathrm{I} \end{pmatrix}, \qquad \hat{\gamma}_k = \begin{pmatrix} 0 & -\hat{\sigma}_k \\ \hat{\sigma}_k & 0 \end{pmatrix},$$

$(\hat{\sigma}_1\hat{\sigma}_2\hat{\sigma}_3 = \sqrt{-1}$ I, where I is the 2×2 unit matrix and $\hat{\sigma}_k$ are the usual 2×2 Pauli matrices)

of a single isospace vector but as four orthonormal basis vectors for real space-time. Stated more explicitly, γ_0 is the unit vector in the forward light cone and γ_k $(k = 1, 2, 3)$ is a dextral set of space-like vectors. The future-pointing timelike vector γ_0 characterizes an observer's rest frame and maps the space-time bivectors $\{\sigma_k\}$ (i.e., $\{\gamma_k\gamma_0\}$) into the orthonormal basis vectors in Pauli algebra. Thus Pauli algebra is identified with the algebra for the rest space relative to the timelike vector γ_0. Also the γ_0 vector determines a map of any space-time vector $a = a^\mu\gamma_\mu$ as

$$a\gamma_0 = a \cdot \gamma_0 + a \wedge \gamma_0. \tag{5.20}$$

Then the scalar part $a \cdot \gamma_0$ is the γ_0-time component of the vector a, and the bivector part $a \wedge \gamma_0$ can be decomposed into the $\{\sigma_k\}$-frame and shown to represent a spatial vector relative to an observer in the γ_0-frame.

The above fact can be visualized as follows: to an observer in the γ_0-frame, any vector appears to be a line segment that exists for some time, so its natural representation in space-time is a bivector. This important and novel feature is embodied by Equation 5.20, which, in fact, demonstrates that the algebraic properties of vectors in relative space, identified in this case by Pauli algebra, are completely determined by those of the relativistic space-time Dirac algebra.

We note that in the transition from Dirac to Pauli algebra the six space-time bivectors $\{\sigma_k, i\sigma_k\}$ with i representing the unit pseudoscalar for Dirac algebra are split up into relative vectors $\{\sigma_k\}$ and relative bivectors $\{i\sigma_k\}$ (i.e., relative to γ_0 observer) with i representing the unit pseudoscalar for Pauli algebra. This split of space-time bivectors is a frame-dependent operation. By this transition from Dirac to Pauli algebra, one can have the observables in a given frame from relativistic quantities in a simple way.

Denoting by $\mathfrak{D}^k$ the subspace of multivectors of grade k in Dirac algebra $\mathfrak{D}$, we may write

$\mathfrak{D}^1 = a$ four-dimensional space of vectors,

$\mathfrak{D}^2 = a$ six-dimensional space of bivectors,

$\mathfrak{D}^3 = a$ four-dimensional space of trivectors (or pseudovectors),

$\mathfrak{D}^4 = a$ one-dimensional space of quadrivectors.

We know that any vector a of space-time algebra anticommutes with i:

$$ai = -ia. \tag{5.21}$$

From (1.30) and (1.31) we can write, by using (5.21),

$$a \wedge i = (1/2)(ai + ia) = 0, \tag{5.22}$$
$$a \cdot i = (1/2)(ai - ia) = ai = -ia. \tag{5.23}$$

$a \cdot i$ or ai is a trivector called the dual of a. Earlier we have seen that in $\mathfrak{D}$ every trivector T is the dual of some vector a:

$$T = ai. \tag{5.24}$$

Multiplying on the right by i ,we obtain from (5.24)

$$Ti = -a \tag{5.25}$$

So, the dual of a trivector T is a unique vector $-a$. This establishes an isomorphism of the linear space $\mathfrak{D}^3$ to the vector space $\mathfrak{D}^1$. So, trivectors of $\mathfrak{D}$ are often called pseudovectors.

It is important to note that in $\mathfrak{D}$, i commutes with bivectors but anticommutes with vectors and trivectors, whereas in Pauli algebra i commutes with vectors and bivectors.

5.3 Conjugations

Following Hestenes [2, 7, 12], we define four types of conjugation operators in real Dirac algebra $\mathfrak{D}$. Any multivector M in $\mathfrak{D}$ can be written as

$$M = M_S + M_V + M_B + M_T + M_P, \tag{5.26}$$

where the subscripts S, V, B, T, and P mean, respectively, scalar, vector, bivector, trivector (pseudovector), and pseudoscalar parts of the multivector M.

5.3.1 Conjugate Multivectors (Reversion)

The conjugate multivector or reversion of M of $\mathfrak{D}$ is denoted by $\tilde{M}$ and is defined by the reversion of the order of products of all vectors of M. In accordance with the notation adopted in current literature, we use the tilde over M to denote the operation of reversion. Thus, the operation of conjugation or reversion takes M into $\tilde{M}$ as

$$\begin{aligned}\tilde{M} &= \tilde{M}_S + \tilde{M}_V + \tilde{M}_B + \tilde{M}_T + \tilde{M}_P \\ &= M_S + M_V - M_B - M_T + M_P.\end{aligned} \tag{5.27}$$

As it is independent of any basis in the algebra, it is an invariant type of conjugation (see also Chapter 2, Section 2.2, Equation 2.17, Equation 2.18, Equation 2.19, and Equation 2.20), and as for pseudoscalar, see Chapter 4, Equation (4.14b).

5.3.2 Space-Time Conjugation

We introduce the space-time conjugation of M of $\mathfrak{D}$ denoted by $\bar{M}$ and defined by

$$\bar{M} = -iMi. \tag{5.28}$$

This operation maps M into $\bar{M}$ as

$$\begin{aligned}\bar{M} &= -iM_Si - iM_Vi - iM_Bi - iM_Ti - iM_Pi \\ &= M_S - M_V + M_B - M_T + M_P.\end{aligned} \tag{5.29}$$

In other words, the space-time conjugation of M is the operation that reverses the direction of all vectors in space-time. From (5.29) we note that the multivector M is even, if $\bar{M} = +M$, and odd, if $\bar{M} = -M$.

5.3.3 Space Conjugation

Next, we introduce the space conjugation of M of $\mathfrak{D}$ denoted by M^* and defined by

$$M^* = \gamma_0 M \gamma_0. \tag{5.30}$$

This operation depends on the choice of γ_0 and changes $\{\sigma_k\}$ and $\{\gamma_k\}$, $(k = 1, 2, 3)$, from right-handed to left-handed frames without affecting γ_0:

$$\sigma_k^* = \gamma_0 \sigma_k \gamma_0 = -\sigma_k, \tag{5.31}$$

$$\gamma_\mu^* = \gamma_0 \gamma_\mu \gamma_0 = (\gamma_0, -\gamma_k). \tag{5.32}$$

As this operation depends on the choice of γ_0, it is not an invariant type of conjugation.

5.3.4 Hermitian Conjugation

Finally, we introduce the Hermitian conjugation of M of $\mathfrak{D}$, conventionally denoted by $M^\dagger$ and defined by

$$M^\dagger = \gamma_0 \tilde{M} \gamma_0. \tag{5.33}$$

The operation $M^\dagger$ corresponds to the Hermitian conjugate of M in the usual matrix representations of the Dirac algebra of matrices. As $M^\dagger$ depends on the choice of γ_0, it is not an invariant type of conjugation.

5.4 Lorentz Rotations

In the algebra of Dirac matrices, the conditions (5.1a, b) do not determine the Dirac matrices uniquely. Any two sets of Dirac matrices $\{\gamma_k\}$ and $\{\gamma_k'\}$ are related by a similarity transformation

$$\gamma_\mu{}' = R\gamma_\mu R^{-1}, \tag{5.34}$$

where R is a nonsingular matrix. This, in fact, gives a change in the representation of the Dirac matrices.

What does Equation 5.34 mean in space-time algebra where the γ_μ are vectors? The geometrical requirement that the γ_μ in (5.34) must be vectors implies that they can be expressed as

$$\gamma_\mu{}' = a_\mu^\nu \gamma_\nu. \tag{5.35}$$

This means that (5.34) must be invariant under reversion

$$(\gamma_\mu' R)^\sim = (R\gamma_\mu)^\sim,$$

that is,

$$\tilde{R}\gamma_\mu{}' = \gamma_\mu \tilde{R}. \tag{5.36}$$

By using (5.34), (5.36) can be written as

$$\tilde{R}R\gamma_\mu R^{-1} = \gamma_\mu R^{-1}. \tag{5.37}$$

So, one may choose R such that

$$\tilde{R}R = 1 \tag{5.38a}$$

giving

$$R^{-1} = \tilde{R}. \tag{5.38b}$$

Then, in view of (5.35) and (5.38b), (5.34) assumes the form

$$\gamma_\mu{}' = a_\mu^\nu \gamma_\nu = R\gamma_\mu \tilde{R}, \tag{5.39}$$

describing a Lorentz transformation from a frame of vectors $\{\gamma_\mu\}$ into a frame $\{\gamma_\mu'\}$. Furthermore one can solve (5.39) for R as a function of γ_μ and γ_μ' only. This implies that R is a multivector and that every Lorentz transformation can be expressed in that form.

Hestenes [1] has shown that (5.39) is a proper Lorentz transformation (i.e., transformations continuously connected to the identity) if and only if R is an even multivector satisfying

$$R\tilde{R} = 1, \tag{5.40}$$

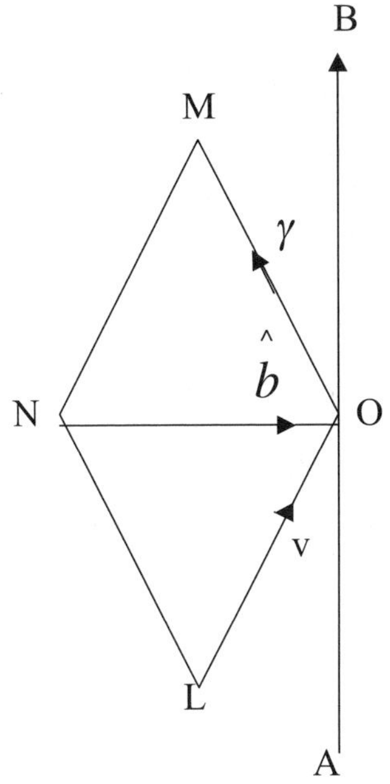

FIGURE 5.1
NO is perpendicular to the hyperplane AOB. v is the incident vector, r is the reflected vector, and $\hat{b}$ is the unit vector along NO. In geometric algebra, r is related to v and $\hat{b}$ by $r = -\hat{b}v\hat{b}$.

from which one can write

$$R = \pm \exp(\varphi/2), \tag{5.41}$$

where φ is a bivector. R is often referred to as a Lorentz rotation.
In fact, R is a 'real' spinor field in space-time algebra.

Now we will show that R can be obtained directly as the geometric product of two unit vectors $\hat{a}$ and $\hat{b}$ that specify the plane of rotation. For this purpose we consider an incident vector v denoted by $\overline{LO}$ reflected along OM by hyperplane AOB perpendicular to the unit vector $\hat{b}$ (see Figure 5.1). Let r be the reflected vector denoted by $\overline{OM}$. As $|v| = |r|$, the diagonal ON of the parallelogram LOMN is perpendicular to the hyperplane AOB. So, we can write

$$\overline{ON} = -2v \cdot \hat{b}\hat{b}. \tag{5.42}$$

Then, the reflected vector r can be written by using (5.42) as

$$\begin{aligned} r = \text{OM} = LN = LO + ON &= v - 2v \cdot \hat{b}\hat{b} \\ &= v - (v\hat{b} + \hat{b}v)\hat{b}. \end{aligned}$$

This simplifies to

$$r = -\hat{b}v\hat{b} \tag{5.43}$$

The relation (5.43) is unique to geometric algebra. Next, we apply a reflection of the vector r in the hyperplane perpendicular to a second unit vector $\hat{a}$. Then, by applying the formula (5.43), the final reflected vector v' can be written as

$$\begin{aligned} v' &= -\hat{a}r\hat{a} \\ &= \hat{a}(\hat{b}v\hat{b})\hat{a} && \text{[by using (5.43)]} \\ &= \hat{a}\hat{b}v\hat{b}\hat{a} \\ &= (\hat{a}\hat{b})v(\hat{a}\hat{b})^{\sim}. \end{aligned} \tag{5.44}$$

The combination of two reflections is a rotation in the plane determined by the reflection axes $\hat{a}$ and $\hat{b}$. So, Equation 5.44 indicates that the vector v is rotated through an angle defined by the directions $\hat{a}$ and $\hat{b}$ to the vector v'. The formulation (5.44) for encoding rotations can be expressed in the compact form:

$$v \Longrightarrow v' = Rv\tilde{R}, \tag{5.45}$$

where

$$R = \hat{a}\hat{b} \tag{5.46}$$

The spinor R defined by (5.46) has the fundamental property

$$R\tilde{R} = \hat{a}\hat{b}\hat{b}\hat{a} = 1. \tag{5.47}$$

From Equation 2.45 and Equation 2.46 we can express (5.46) in the form

$$R = e^{i\vartheta} = \cos\vartheta + i\sin\vartheta, \tag{5.48}$$

where the angle ϑ is expressed in radians. Equation 5.48 shows that R is an even multivector or spinor consisting of scalar and bivector parts.

By using (2.50) we can express (5.48) as

$$R = \pm e^{\varphi/2}, \tag{5.49}$$

where φ is a bivector given by

$$\varphi = i\vartheta. \tag{5.50}$$

Here $\varphi/2$ encodes the areal magnitude of the angle ϑ (in radian measure) from a to b (see Figure 2.3 and Figure 2.4). The sign of ϑ in (5.50) is determined by the orientation assigned to the unit pseudoscalar i as shown in Figure 2.3 and Figure 2.4. Equation 5.49 is a natural generalization of the complex representation used in two dimensions.

We see that the spinor R, which is often referred to as a Lorentz rotation, is given by

$$R = \exp(\varphi/2), \tag{5.51}$$

where the bivector φ is expressed as a linear combination of six linearly independent bivectors $\{\sigma_k, i\sigma_k\}$. So, we can write

$$R = \exp(\mu/2 + i\lambda/2), \tag{5.52}$$

where μ and λ are bivectors satisfying

$$\mu^* = -\mu, \tag{5.53a}$$

$$\lambda^* = -\lambda. \tag{5.53b}$$

Then, (5.52) can be expressed as

$$R = R_2 R_1, \tag{5.54}$$

where

$$R_1 = \exp(i\lambda/2), \tag{5.55a}$$

$$R_2 = \exp(\mu/2), \tag{5.55b}$$

such that

$$R_1 = R_1^*, \tag{5.56a}$$

$$R_2 = R_2^* \tag{5.56b}$$

The spinor R_1 in (5.55a) satisfying the condition (5.56a) is called a spatial rotation, whereas the spinor R_2 in (5.55b) satisfying the condition (5.56b) is called a special timelike rotation (see References [2,7,14]). Equation 5.52 implies that any Lorentz rotation can be expressed as a spatial rotation followed by a special timelike rotation.

In Pauli algebra the space-time bivectors μ and λ become vectors, and $i\lambda$ is the bivector with respect to the γ_0-frame, i being the unit pseudoscalar for Pauli algebra. So, the spinor R in Pauli algebra can be expressed as

$$R = \exp(i\lambda/2). \tag{5.57}$$

5.5 Spinor Theory of Rotations in Three-Dimensional Euclidean Space

As already stated in Section 5.2, that Equation 5.20 demonstrates that the algebraic properties of vectors in the relative space of Pauli algebra (i.e., three-dimensional Euclidean space) are completely determined by those of space-time algebra, we are in a position to develop the general theory of

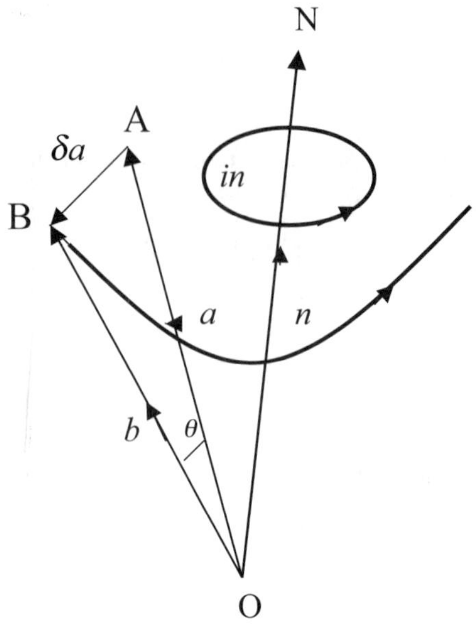

FIGURE 5.2
The schematic diagram of a rigid body rotating about a point O with the paths of all points of it parallel to a plane represented by a bivector in, where n is the unit vector that specifies the rotation axis ON. ϑ is an angle of infinitesimal rotation from point A to point B, where $\overline{OA} = a$ and $\overline{OB} = b$. Then, we have $\delta a = b - a = i(n \wedge a)(\vartheta)$ up to the first order of ϑ.

rotations in three-dimensional Euclidean space by means of the Lorentz rotation R, which is a real spinor field in space-time algebra.

In Section 5.4 we have seen that the spinor R in Pauli algebra can be expressed as

$$R = \exp(i\lambda/2), \tag{5.58}$$

where λ is a vector and i is the unit pseudoscalar for Pauli algebra. So, $i\lambda$ is a bivector.

Now we consider a rigid body rotating about a reference point O with the paths of all points of it parallel to some plane. We can represent this plane by a bivector of unit modulus in, where n is the unit vector that specifies the rotation axis ON (see Figure 5.2).

Then, the bivector $i\lambda$ in Equation 5.58 can be expressed as [11, 12]:

$$i\lambda = in\vartheta, \tag{5.59}$$

where ϑ is the angle of rotation about the n-direction. So, the spinor R in (5.58) can be written as

$$R = \exp(in\vartheta/2) = \cos(\vartheta/2) + in\sin(\vartheta/2), \tag{5.60}$$

which satisfies the condition (5.40).

The action of the spinor R on a vector a is written as (see Equation 5.39)

$$b = Ra\tilde{R}. \tag{5.61}$$

Thus, the vector a in relative space is rotated through an angle ϑ to the vector b by the above action of R on a. In view of Equation 5.60 and Equation 5.61, we notice that the vector b can be obtained via the geometric product of the vector a and unit vector n as

$$\begin{aligned} b &= \exp(in\vartheta/2)a\exp(-in\vartheta/2) \\ &= [\cos(\vartheta/2) + in\sin(\vartheta/2)]a[\cos(\vartheta/2) - in\sin(\vartheta/2)]. \end{aligned} \tag{5.62}$$

For infinitesimal rotation, we have from (5.62), retaining terms only up to the first order of ϑ,

$$\begin{aligned} b &= [1 + (in\vartheta/2)]a[1 - (in\vartheta/2)] \\ &= a + (in\vartheta/2)a - a(in\vartheta/2) \\ &= a + i(na - an)(\vartheta/2) \\ &= a + i(n \wedge a)\vartheta. \end{aligned} \tag{5.63}$$

In the above computation we note that in Pauli algebra i commutes with vectors, indicating its similarity in character with the unit imaginary. Thus, we have from (5.63) the variation of the vector a up to the first order of ϑ:

$$\delta a = (b - a) = i(n \wedge a)\vartheta, \tag{5.64}$$

which shows that δa is orthogonal to both n and a.

We have seen from Equation 5.61 that by the action of the spinor R on the vector a in Pauli algebra, the vector a is rotated through an angle ϑ to the vector b corresponding to the spatial rotation of $\vartheta/2$ of the spinor in the 'real' phase space. So, spatial rotation R can be identified with the generalized phase φ of a neutron spin state (an abstract space) of quantum theory. Mathematically, this is represented by

$$\text{rotation angle } \vartheta \text{ of the vector } = 2\varphi(\text{mod } 360^0), \tag{5.65}$$

where φ is the generalized phase angle (spatial rotation). This, in fact, demonstrates the well-known change of sign of a fermion spin for rotation of 2π in quantum theory whose mathematical counterpart is the one-sidedness of the Möbius strip.

For further elucidation of the above discussion we consider the composition of rotations. If we perform a rotation of the vector b with another spinor R', we obtain a vector c given by

$$c = R'b\tilde{R}' = (R'R)a(\tilde{R}\tilde{R}') = (R'R)a(R'R)^{\tilde{}}. \tag{5.66}$$

Here we use Equation 5.61 and note that the geometric product is associative. Equation 5.66 implies that the composition $R'R$ of rotations is also a rotation that corresponds to the rotation from the vector a to the vector c.

It represents the law of left composition of rotation group such that R' operates on R from the left without touching the vector a.

For a rotation of the vector b through an angle 2π, the spinor R' becomes

$$R' = \tilde{R}' = -1, \tag{5.67a}$$

which gives

$$R'R = -R. \tag{5.67b}$$

Equation 5.66 in this case turns out to be

$$c = b = (-R)a(-\tilde{R}), \tag{5.68}$$

The first part of Equation 5.68 means that the vector b remains unchanged for a rotation through an angle 2π under the action of R', whereas the second part, together with Equation 5.61, tells us that either R or $-R$ rotates the vector a to the vector b, i.e., both represent the same rotation. We know that R and $-R$ are two distinct elements of the geometric algebra. So, the correspondence between spinors and rotations is 2 to 1. This is the well-known relation between the matrices of the unitary group $SU(2)$ and the matrices of the orthogonal group SO(3).

Thus, we note that the well-known change of sign of a fermion spin for a rotation of 2π in quantum theory is due to the fact that the spinor R' acts on one side of the spinor R (see Equation 5.67a, b). On the other hand, for a vector, as shown previously (see Equation 5.68), the spinor R' acts on both sides and thus exibits no change of sign of the vector for a rotation of 2π under the action of R'.

Later we will show that the rotation angle ϑ, described by the vector a about the n-direction, can be related via spatial rotation R in (5.68) to the generalized phase of a neutron spin state represented by some point in the total space of the fiber bundle of a neutron spin rotation.

References

1. D. Hestenes, *Space-Time Algebra* (Gordon and Breach, New York, 1967).
2. D. Hestenes, *J. Math. Phys.* 14, 893 (1973).
3. D. Hestenes, *J. Math. Phys.* 16, 556 (1975).
4. D. Hestenes and R. Gunther, *J. Math. Phys.* 16, 573 (1975).
5. D. Hestenes, *Am. J. Phys.* 47, 399 (1979).
6. D. Hestenes, *Am. J. Phys.* 39, 1013 (1971).
7. D. Hestenes and G. Sobczyck, *Clifford Algebra to Geometric Calculus* (Reidel, Boston, 1984).
8. J. D. Hamilton, *J. Math. Phys.* 25, 1823 (1984).
9. D. Hestenes, *New Foundations for Classical Mechanics* (Reidel, Boston, 1986).

10. B. K. Datta, "Physical theories in space-time algebra," in *Quantum Gravity*, eds. P. G. Bergmann, V. de Sabbata, and H. J. Treder (World Scientific, Singapore, 1996), pp. 54–79.
11. B. K. Datta, V. de Sabbata, and L. Ronchetti, *Nuovo Cimento B*, 113, 711 (1998).
12. B. K. Datta and V. de Sabbata, "Hestenes' geometric algebra and real spinor fields," in *Spin in Gravity: Is It Possible to Give an Experimental Basis to Torsion*? eds. P. G. Bergmann, V. de Sabbata, G. T. Gillies, and P. I. Pronin (World Scientific, Singapore, 1998), pp. 33–50.
13. B. K. Datta, R. Datta, and V. de Sabbata, *Found. Phys. Letters* 11, 83 (1998).
14. V. de Sabbata, L. Ronchetti, and B. K. Datta, "Non-commuting geometry and spin fluctuations," in *Classical and Quantum Non-locality*, eds. P. G. Bergmann, V. de Sabbata, and J. N. Goldberg (World Scientific Singapore, 2000), pp. 85–110.

6

Spinor and Quaternion Algebra

6.1 Spinor Algebra: Quaternion Algebra

In Section 3.5 of Chapter 3, we have shown that the algebra of complex numbers appears with a geometric intepretation as the even subalgebra $\mathcal{G}_2^+$ of $\mathcal{G}_2$. In the same way, we can express $\mathcal{G}_3$ as the sum of an odd multivector part $\mathcal{G}_3^-$ and an even multivector part $\mathcal{G}_3^+$:

$$\mathcal{G}_3 = \mathcal{G}_3^- + \mathcal{G}_3^+. \tag{6.1}$$

Then, it follows from Equation 4.13 that a multivector M in $\mathcal{G}_3$ can be put in the form:

$$M = M^- + M^+, \tag{6.2}$$

where

$$M^- = a + i\beta, \tag{6.3}$$

$$M^+ = \alpha + ib. \tag{6.4}$$

One can show that $\mathcal{G}_3^+$ is closed under multiplication, so it is a subalgebra of $\mathcal{G}_3$, but $\mathcal{G}_3^-$ is not. This even subalgebra $\mathcal{G}_3^+$ may plausibly be called spinor algebra to emphasize the geometric significance of its elements. In Section 3.4 of Chapter 3 we have shown that every spinor in $\mathcal{G}_2^+$ represents a rotation–dilation in two-dimensional Euclidean plane. In the same way, one can show that every spinor in $\mathcal{G}_3^+$ represents a rotation–dilation in three-dimensional Euclidean space E_3, which represents a subspace $\mathcal{G}_3^1$ of vectors in $\mathcal{G}_3$.

In what follows, we shall show that the spinor algebra $\mathcal{G}_3^+$ is isomorphic to the quaternion algebra developed by William Rowan Hamilton in 1843.

For this purpose we express Equation 6.4, using Equation 4.7a, b, c, Equation 4.9 and Equation 4.11, in the form

$$M^+ = \alpha 1 + \mathcal{B}_1 i_1 + \mathcal{B}_2 i_2 + \mathcal{B}_3 i_3. \tag{6.5}$$

This shows that the set of four unit multivectors $\{1, i_1, i_2, i_3\}$ makes up a basis for $\mathcal{G}_3^+$. Thus, $\mathcal{G}_3^+$ is a linear space of four dimensions. The elements of $\mathcal{G}_3^+$ were first introduced and called quaternions by Hamilton. So, the spinor algebra $\mathcal{G}_3^+$ and the quaternion algebra of Hamilton are identical.

Quaternions and spinors have equivalent algebraic properties as well as the same geometric significance. So, quaternions are spinors. Hamilton, in fact, found a way to describe geometry by algebra by introducing a system of quantities to represent rotations in three dimensions by generalizing the concept of complex numbers.

In this connection a brief digression of quaternion algebra will be in order. W. R. Hamilton developed the concept of directed numbers to emphasize the operational interpretation, and constructed quaternions in 1843 by generalizing the concept of complex numbers. By introducing a system of imaginary units $\hat{\imath}, \hat{\jmath}, \hat{k}$ to represent rotations in three-dimensional Euclidean space, he defined a quaternion q as

$$q = \alpha + \hat{\imath}x + \hat{\jmath}y + \hat{k}z \tag{6.6}$$

(here "ˆ" indicates the quaternion units) where α, x, y, z are scalars, and $\hat{\imath}, \hat{\jmath}, \hat{k}$ satisfy the following defining relations:

$$\hat{\imath}^2 = \hat{\jmath}^2 = \hat{k}^2 = \hat{\imath}\hat{\jmath}\hat{k} = -1. \tag{6.7}$$

The set of four quantities $\{1, \hat{\imath}, \hat{\jmath}, \hat{k}\}$ makes up a basis for the quaternion algebra.

The quaternion q defined above can also be described in terms of the basis $\{I, j\hat{\sigma}_1, j\hat{\sigma}_2, j\hat{\sigma}_3\}$, where I is the 2×2 identity matrix, j is the unit imaginary $\sqrt{-1}$, and $\hat{\sigma}_k$ $(k = 1, 2, 3)$ are the Pauli matrices. This formulation of the quaternions in the guise of Pauli matrices provides a great resurgence of the quaternions depicting their essential role in the quantum theory of spin and quantum field theory.

Moreover, the preceding discussion shows that quaternion algebra is not only isomorphic with spinor algebra $\mathcal{G}_3^+$ but also identical with it. In fact, both have equivalent algebraic properties as well as the same geometric significance. So, quaternions are spinors.

The embedding of quaternion algebra in the geometric algebra of three-dimensional Euclidean space exhibits that quaternions are bivectors (see Equation 4.7a, b, c). Thus, the roles of vectors and quaternions (bivectors) are explicitly distinguished because vector algebra and quaternion algebra are subalgebras $\mathcal{G}_3^1$ and $\mathcal{G}_3^+$ of geometric algebra $\mathcal{G}_3$. So, the two systems complement each other.

We note in passing that the bivectors $\{i_1, i_2, i_3\}$ in Equation 6.5 of spinor algebra satisfy equations

$$i_1^2 = i_2^2 = i_3^2 = -1$$

and

$$\begin{aligned} i_1 i_2 i_3 = i\sigma_1 i\sigma_2 i\sigma_3 &= i\sigma_1\sigma_2 i i\sigma_3 = -i_3\sigma_1\sigma_2\sigma_3 \\ &= -ii = 1. \end{aligned}$$

Thus, we have

$$i_1^2 = i_2^2 = i_3^2 = -i_1 i_2 i_3 = -1. \tag{6.8}$$

The difference in sign between $\hat{\imath}\ \hat{\jmath}\ \hat{k}$(of Hamilton) and $i_1\ i_2\ i_3$ (of Hestenes) can be done away by making the following correspondence:

$$\begin{aligned} \hat{\imath} &= i_1 = i\sigma_1 \\ \hat{\jmath} &= -i_2 = -i\sigma_2 \\ \hat{k} &= i_3 = i\sigma_3. \end{aligned} \tag{6.9}$$

6.2 Vector Algebra

Now we show that vector algebra as developed by J. Willard Gibbs in 1884 fits naturally into $\mathcal{G}_3$. For this purpose we first obtain the relation between the outer product $a \wedge b$ and the cross product $a \times b$ as introduced by Gibbs.

From (4.7c) we can write

$$\alpha\sigma_1 \wedge \beta\sigma_2 = i\alpha\beta\sigma_3, \tag{6.10}$$

where α, β are scalars. This implies that the bivector $\alpha\sigma_1 \wedge \beta\sigma_2$ is the dual of the vector $\alpha\beta\sigma_3$, which is perpendicular to the oriented rectangle described by the bivector. Moreover,

$$|\alpha\sigma_1 \wedge \beta\sigma_2| = |\alpha\beta\sigma_3|. \tag{6.11}$$

This is well exemplified by the following relation between the outer product $a \wedge b$ and the cross product $a \times b$ [1–3]:

$$a \wedge b = ia \times b \tag{6.12}$$

or equivalently,

$$a \times b = -ia \wedge b. \tag{6.13}$$

In Figure 6.1 the sign of the duality is chosen such that the vectors a, b, $a \times b$ form a dextral set in agreement with our convention for the handedness of the dextral pseudoscalar i . This is evident from (4.7c) and (6.10) as shown below:

$$\sigma_1 \wedge \sigma_2 = i\sigma_3 \Longrightarrow i\sigma_1 \times \sigma_2 = i\sigma_3 \Longrightarrow \sigma_1 \times \sigma_2 = \sigma_3. \tag{6.14}$$

Finally, we note from (6.12) or (6.13) that the magnitude of the vector $a \times b$ is equal to the area of the parallelogram in Figure 6.1, which can be identified with $|\ a \wedge b\ |$.

In Gibbs' vector algebra, the cross product $a \times b$ is termed an axial vector if a and b are polar vectors. In geometric algebra, an axial vector appears as a bivector disguised as the dual of a vector. According to the relation (6.12), the

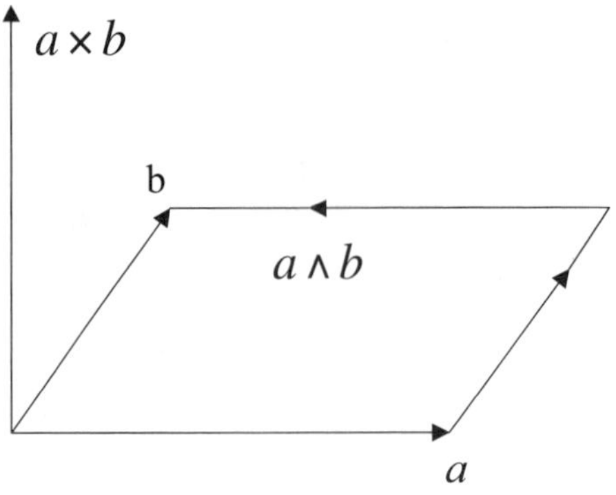

FIGURE 6.1
Duality of the cross product and the outer product.

cross product $a \times b$ is a vector in exactly the same sense as a and b are vectors, which removes the so-called distinction between axial and polar vectors. So, we see that Gibbs' vector algebra fits naturally into $\mathcal{G}_3$.

6.3 Clifford Algebra: Grand Synthesis of Algebra of Grassmann and Hamilton and the Geometric Algebra of Hestenes

In an ingenious way William Rowan Hamilton described quaternion algebra by introducing a system of quantities to represent rotations in three-dimensional Euclidean space. In fact, he developed the concept of directed numbers to emphasize the operational interpretation, and constructed quaternions in 1843 by generalizing the concept of complex numbers. We have seen in the previous section that quaternion algebra is identical with spinor algebra $\mathcal{G}_3^+$, which is an even subalgebra of geometric algebra $\mathcal{G}_3$. So, quaternion are spinors. It is well known that quaternions have reappeared in the guise of Pauli matrices and played an important role in describing quantum mechanics and quantum field theory.

One year later, in 1844, Herman Grassmann developed his "algebra of extension" based on the inner and outer products of vectors. In fact, he developed the concept of directed numbers from the quantitative point of view. In his later life, Grassmann realized that Hamilton algebra was related to his, and quaternions could be derived simply by the introduction of the geometric product ab of vectors defined by $ab = a \cdot b + a \wedge b$. However, it was too late for him to pursue the far-reaching implications of the grand mixture of geometric and algebraic ideas.

Soon afterwards the English mathematician William Kingdon Clifford took up the problem as realized by Grassmann. By virtue of his powerful geometric insight he realized that both Grassmann and Hamilton were developing the same subject from different points of view; Grassmann developed the concept of directed numbers from the quantitative standpoint, whereas Hamilton developed it with emphasis on the operational interpretation. In 1876, Clifford synthesized the algebraic ideas of Grassmann and Hamilton by developing the concept of the geometric product of vectors and was able to construct the structure of a general algebraic system now known as Clifford algebra. Unfortunately for the mathematical world, death claimed his life before he could complete the delineation of the potential combination of geometric and algebraic ideas.

After about nine decades, David Hestenes [1–3] developed geometric algebra during the decades 1966–86, by combining the algebraic structure of Clifford algebra with the explicit geometric meaning of its mathematical elements at its foundation, and accomplished fully the delineation of the grand combination of geometric and algebraic ideas. So, formally, it is Clifford algebra endowed with geometric information of and physical interpretation to all mathematical elements of the algebra. As it stands, it is the largest possible associative division algebra that integrates all algebraic systems (i.e., algebra of complex numbers, vector algebra, matrix algebra, quaternion algebra, etc.) into a coherent mathematical language that augments the powerful geometric intuition of the human mind with the precision of an algebraic system. Hestenes first signified that two different interpretation of numbers could be distinguished: quantitative and operational. In Chapter 3, Sections 3.2, 3.3 and 3.4, this distinction is explicitly illustrated by the interpretations given to a unit bivector i, called the unit pseudoscalar for the i-plane. Interpreted quantitatively, i is a measure of a directed area, and operationally interpreted, i signifies a rotation in the i-plane. Either a quantitative or an operational interpretation can be given to any number, called a multivector in geometric algebra. Thus, vectors are usually interpreted quantitatively, whereas spinors, which are even multivectors, are usually interpreted operationally.

A lively controversy arose in the last two decades of the 19th century as to which system was more suitable for the development of theoretical physics: quaternion formulation due to Hamilton or the newer vector algebra of Josiah Willard Gibbs and Oliver Heaviside. As vectors were tailored by Gibbs and Heaviside independently to each other, to formulate the electromagnetic theory in an elegant way, they proved to be easier and more useful than quaternions. Nevertheless, the reappearance of the quaternions in the guise of Pauli matrices in the 20th century provides a great resurgence of the quaternions, depicting their essential role in quantum mechanics and quantum field theory.

Geometric algebra explicitly exhibits that the statement

vectors versus quaternions

is devoid of any meaning. In the foregoing discussion it is shown that Gibbs' vector algebra and Hamilton's quaternion algebra are, respectively, the subalgebras $\mathcal{G}_3$ and $\mathcal{G}_3^+$ of geometric algebra $\mathcal{G}_3$. Thus, the two systems complement each other, both being united in geometric algebra $\mathcal{G}_3$, and the vector–quaternion controversy recedes into the shadows.

References

1. D. Hestenes, *Space-Time Algebra* (Gordon and Breach, New York, 1967).
2. D. Hestenes and G. Sobczyck, *Clifford Algebra to Geometric Calculus* (Reidel, Boston, 1984).
3. D. Hestenes, *New Foundations for Classical Mechanics* (Reidel, Boston, 1986).

Part II

7

Maxwell Equations

7.1 Maxwell Equations in Minkowski Space-Time

As an example, we like to write the Maxwell equations in the formalism of geometric algebra. Start with the Maxwell equations written in Gauss units (where the electric field E, the charge ρ, and the current density j are expressed in electrostatic unit [esu] or [CGS] esu, whereas the magnetic field is measured in electromagnetic unit [emu] or [CGS] emu):

$$\begin{cases} rot E + (1/c)(\partial H/\partial t) = 0 & (7.1) \\ div H = 0 & (7.2) \end{cases}$$

$$\begin{cases} rot H - (1/c)(\partial E/\partial t) = (4\pi/c)\, j & (7.3) \\ div E = 4\pi\rho & (7.4) \end{cases}$$

or

$$\nabla \times E + (1/c)\dot{H} = 0 \tag{7.5}$$

$$\nabla \cdot H = 0 \tag{7.6}$$

$$\nabla \times H - (1/c)\dot{E} = (4\pi/c)\, j \tag{7.7}$$

$$\nabla \cdot E = 4\pi\rho \tag{7.8}$$

where

$$\begin{aligned}
&\nabla \Longrightarrow (\partial/\partial x) + (\partial/\partial y) + (\partial/\partial z)\\
&\nabla\varphi = \text{grad}\,\varphi\\
&\nabla \cdot A = \text{div } A\\
&\nabla \times A = \text{rot } A\\
&\nabla \cdot \nabla \times A = \text{div rot } A = 0\\
&\nabla \times \nabla\varphi = \text{rot grad}\,\varphi = 0\\
&\nabla \times (\nabla \times A) = (\nabla \cdot A)\nabla - (\nabla \cdot \nabla) A \Longrightarrow \text{rot rot } A\\
&\qquad = \text{grad div A} - \text{div grad A, etc.}
\end{aligned}$$

Put

$$H = rot\,A \tag{7.9}$$

Substituting (7.9) in (7.1) we have

$$E + (1/c)(\partial A/\partial t) = -\text{grad}\,V \tag{7.10}$$

$$\text{i.e.,}\quad E = -\text{grad}\,V - (1/c)(\partial A/\partial t) \tag{7.11}$$

The vector potential A and the scalar potential V are not uniquely defined. In fact, the vector potential is determined to within the gradient of an arbitrary function φ, and the scalar potential to within the time derivative of the same function φ; that is, if instead of A and V we take A_o and V_o, where

$$A_o \Longrightarrow A + grad\,\varphi;\quad V_o \Longrightarrow V - (1/c)(\partial\varphi/\partial t), \tag{7.12}$$

we see that A_o and V_o satisfy Equation 7.9 and Equation 7.11;

$$\text{i.e.,}\quad H = \text{rot}\,A_o;\quad E = -\text{grad}\,V_o - (1/c)(\partial A_o/\partial t) \tag{7.13}$$

The fields are invariant with respect to the transformation (7.12) of the potentials; this invariance is called "gauge invariance."

This nonuniqueness of the potentials allows us to choose V in a way that the so-called Lorentz condition (a gauge condition)

$$\text{div}\,A + (1/c)(\partial V/\partial t) = 0 \tag{7.14}$$

be satisfied. Moreover, we have the continuity equation

$$\text{div}\,j + \partial\rho/\partial t = 0 \tag{7.15}$$

In Minkowski space-time, as it is well known, one introduces the tetrapotential

$$A_x \Longrightarrow \phi_1,\quad A_y \Longrightarrow \phi_2,\quad A_z \Longrightarrow \phi_3,\quad iV \Longrightarrow \phi_4. \tag{7.16}$$

Then, the electromagnetic field tensor is

$$F_{\mu\nu} = (\partial\phi_\nu/\partial x^\mu) - (\partial\phi_\mu/\partial x^\nu) \tag{7.17}$$

where

$$F_{\mu\nu}\quad \begin{vmatrix} 0 & H_z & -H_y & -iE_x \\ -H_z & 0 & H_x & -iE_y \\ H_y & -H_x & 0 & -iE_z \\ iE_x & iE_y & iE_z & 0 \end{vmatrix} \tag{7.18}$$

The first pair of Maxwell Equations 7.1 and 7.2 becomes

$$(\partial F_{\mu\nu}/\partial x^{\lambda}) + (\partial F_{\lambda\mu}/\partial x^{\nu}) + (\partial F_{\nu\lambda}/\partial x^{\mu}) = 0 \qquad (7.19)$$

(with $\mu, \nu, \lambda = (1,2,3), (2,3,4), (1,3,4), (1,2,4)$)
and the second pair of Maxwell Equations 7.3 and 7.4

$$\partial F_{\mu\nu}/\partial x^{\nu} = (4\pi/c)\, j_{\mu} \qquad (7.20)$$

(where $j_{\mu} \equiv (j_1, j_2, j_3, ic\rho)$ is the four-dimensional current vector). The equation for the tetra-potential ϕ_{μ} is

$$\Box \phi_{\mu} = -(4\pi/c)\, j^{\mu} \qquad (7.21)$$

with the Lorentz condition

$$\partial \phi_{\mu}/\partial x^{\mu} = 0 \qquad (7.22)$$

where

$$\Box \equiv (\partial^2/\partial x^2) + (\partial^2/\partial y^2) + (\partial^2/\partial z^2) - (1/c^2)(\partial^2/\partial t^2) \qquad (7.23)$$

7.2 Maxwell Equations in Riemann Space-Time (V_4 Manifold)

If we introduce the coupling between electromagnetic and gravitational fields according to the minimal coupling procedure (i.e., replacing partial derivatives with covariant ones, and the flat Minkowski metric with the metric in presence of the gravitational field: $\partial \Longrightarrow \nabla$ and $\eta_{\mu\nu} \Longrightarrow g_{\mu\nu}$), then the first couple of Maxwell equations is unchanged [1], i.e.,

$$(\partial F_{\mu\nu}/\partial x^{\lambda}) + (\partial F_{\lambda\mu}/\partial x)^{\nu} + (\partial F_{\nu\lambda}/\partial x^{\mu}) = 0 \qquad (7.24)$$

due to the identity

$$\partial_{[\mu} F_{\alpha\beta]} = \partial_{[\mu}\partial_{\alpha}\phi_{\beta]} = 0 \qquad (7.25)$$

whereas the second pair becomes

$$F^{\mu\nu}_{;\nu} = (4\pi/c)\, j'^{\mu} \qquad (7.26)$$

where

$$j'^{\mu} = j^{\mu}/\sqrt{-g} \qquad (7.27)$$

(semi-colon ";" stands for covariant derivative).

7.3 Maxwell Equations in Riemann–Cartan Space-Time (U_4 Manifold)

If we introduce torsion in the Einstein general theory of relativity (a problem that was primarily considered by Cartan, and that, in physical sense, means the introduction of spin in gravity theory [2,3,4,5]), we can try to write Maxwell equations in the Einstein–Cartan space-time (U_4 manifold).

In general, one says that the electromagnetic field cannot be coupled to torsion in order to preserve local gauge invariance; in other words, one cannot apply the minimal coupling principle

$$\eta_{\mu\nu} \Longrightarrow g_{\mu\nu}, \quad \partial_\mu \Longrightarrow \nabla_\mu \tag{7.28}$$

that is, one cannot replace the flat with curved metric and the partial derivative with covariant derivative in the presence of torsion to the special relativistic photon Lagrangian without breaking the gauge invariance of the theory.

This result finds a natural justification in the framework of the Poincaré gauge field theory of gravitation (see Reference [6] for an extensive bibliography on this subject).

Several attempts have been made in order to apply minimal coupling without breaking gauge invariance; they are successful, however, only at the cost of imposing arbitrary geometrical constraints upon torsion [7], or introducing a modified definition of gauge transformation [8].

Therefore, these attempts lie beyond the limits of the Einstein–Cartan theory.

In order to preserve the gauge invariance in the Einstein theory, the most simple hypothesis is to assume that ([9,10,11])

photons neither produce nor feel torsion (7.29)

i.e., that Maxwell equations in a Riemann–Cartan space U_4 are identical to the same equations written in the Riemann space V_4 obtained from U_4 putting torsion to zero.

We like to emphasize that the statement (7.29) can be assumed to be strictly valid only as long as the electromagnetic (e.m.) field is treated as a classical, not quantized, field. When a quantum point of view is adopted, however, (7.29) is recovered only as a first approximation because, in general, according to a quantum description of the e.m. field in a space with torsion, one may always expect an interaction between the photons and the torsionic background.

In fact, a photon, with a process of the second order in the perturbative development of the e.m. interaction, can virtually disintegrate into an electron–positron pair (vacuum polarization effect); because these particles are massive fermions, which couple to torsion, they feel the presence of a torsionic background. As a consequence, the e.m. field is also affected by torsion (see Figure 7.1); even if torsion does not directly interact with the photon field, it does interact with the virtual pairs produced in vacuum by a

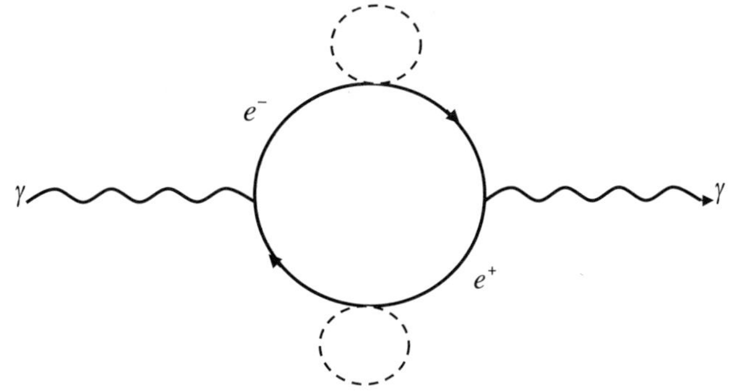

FIGURE 7.1
The dashed lines represent the interaction of the virtual pair with the external torsionic background.

"physical" photon. This interaction preserves the gauge invariance, and the Maxwell equations are modified by a quantum contribution of the second order, so that, to the zeroth order, i.e., in the classical field approximation, the coupling with torsion disappears and we recover the usual form as the Maxwell equations.

We will not go through the calculation (see [12–18]); we will note only that the virtual pair production induces in vacuum a current density proportional to the field intensity. The resulting Maxwell equations are, then, in the presence of torsion (see Reference [15]):

$$\left.\begin{aligned} \partial_\mu F^{\nu\mu} &= 4\pi j^\nu + (2\alpha/3\pi)\eta^{\nu\mu\rho\sigma}F_{\mu\rho}Q_\sigma. \\ \partial_{[\nu}F_{\sigma\mu]} &= 0. \end{aligned}\right\} \tag{7.30}$$

The Equation 7.30 can be derived from a Lagrangian where the photon–torsion interaction term $\mathcal{L}_I$ is given by ([16, 17, 18])

$$\mathcal{L}_I = (-g)^{1/2}\alpha\eta^{\mu\nu\alpha\beta}\phi_\mu F_{\nu\alpha}Q_\beta, \tag{7.31}$$

with $F_{\nu\mu} = 2\partial_{[\nu}A_{\mu]}$. α is the fine-structure constant $\alpha = e^2/\hbar c$ (remember that the fine-structure constant enters in the photon–torsion interaction Lagrangian when an explicit computation is carried out that takes into account the interaction between torsion and the virtual pairs $e^+ - e^-$associated to a propagating electromagnetic field; see, for instance, References [15, 16, 17], and [1] on page 268) and Q_σ is an axial vector related to the torsion tensor $Q^\rho_{\mu\nu} = \Gamma^\rho_{[\mu\nu]}$ by

$$Q^\sigma = (1/16)\eta^{\mu\nu\rho\sigma}K_{\mu\nu\rho} = (1/16)\eta^{\mu\nu\rho\sigma}(Q_{\nu\rho\mu} - Q_{\mu\nu\rho} - Q_{\rho\mu\nu}), \tag{7.32}$$

where $\eta^{\mu\nu\rho\sigma}$ is the totally antisymmetric tensor. This lowest-order photon–torsion interaction has been obtained through an explicit perturbative calculation.

In the case of propagating torsion we can write (see Reference [16]):

$$F^{\mu\nu}_{;\nu} = 4\pi j^{\mu} + (1/8\pi)\alpha\eta^{\mu\nu\alpha\beta} F_{\nu\alpha}\varphi_{,\beta} \tag{7.33}$$

where $\varphi_{,\mu} = (16/3)Q_{\mu}$.

7.4 Maxwell Equations in Terms of Space-Time Algebra (STA)

Now we would like to discuss the Maxwell equations in terms of geometric algebra formalism.

We have seen in Chapter 5 that every multivector M can be written as a linear combination "over the reals" of the 16 elements of STA, i.e.,

$$\begin{matrix} 1 & \gamma_{\mu} & \{\sigma_k, i\sigma_k\} & i\gamma_{\mu} & i \\ 1-scalar & 4-vectors & 6-bivectors & 4-pseudovectors & 1-pseudoscalar \\ grade\ 0 & grade\ 1 & grade\ 2 & grade\ 3 & grade\ 4 \end{matrix} \tag{7.34}$$

where $\sigma_k \equiv \gamma_k\gamma_0$, $k = 1, 2, 3$, and the unit pseudoscalar of space-time is

$$i \equiv \gamma_0 \wedge \gamma_1 \wedge \gamma_2 \wedge \gamma_3 = \gamma_0\gamma_1\gamma_2\gamma_3 = \sigma_1\sigma_2\sigma_3 \tag{7.35}$$

and

$$\gamma_0 i = \gamma_1\gamma_2\gamma_3 \tag{7.36}$$

Then, the multivector M can be written (putting in evidence the parts with different 'grade') as

$$M = \alpha + a + B + bi + \beta i \tag{7.37}$$

where α and β are scalars, a and b are vectors, and B is a bivector:

$$B = (1/2)B^{\mu\nu}\gamma_{\mu} \wedge \gamma_{\nu} \tag{7.38}$$

In order to facilitate the decomposition of a multivector with respect to the basis $\{\gamma_{\mu}\}$, it is convenient to introduce a reciprocal basis system $\{\gamma^{\mu}\}$ defined, as usual, by the conditions

$$\gamma^{\mu} \cdot \gamma_{\nu} = \delta^{\mu}_{\nu} \tag{7.39}$$

This implies

$$\gamma^0 = \gamma_0 \quad \text{and} \quad \gamma^k = -\gamma_k, \tag{7.40}$$

from which it comes out that in terms of that basis (with signature (+ - - -)) we have

$$\gamma_0^2 = -\gamma_k^2 = 1 \quad (k = 1, 2, 3) \quad \text{and} \quad \gamma_{\mu} \cdot \gamma_{\nu} = 0 \quad (\mu \neq \nu) \tag{7.41}$$

that can be assembled in the unique relation

$$g_{\mu\nu} = (1/2)(\gamma_\mu \gamma_\nu + \gamma_\nu \gamma_\mu), \tag{7.42}$$

where $g_{\mu\nu} \equiv \gamma_\mu \cdot \gamma_\nu$ coincides with the Minkowski metric tensor.

For instance, the components of a vector a are the scalar quantities $a^\mu = a \cdot \gamma^\mu$, and then $a = a^\mu \gamma_\mu$. These correspond to the usual controvariant components, whereas the covariant components are given by $a_\mu = a \cdot \gamma_\mu = a^\nu g_{\nu\mu}$. Analogously, the components $B^{\mu\nu} = -B^{\nu\mu}$ of a bivector are (see Equation 7.38):

$$B^{\mu\nu} = (\gamma^\mu \wedge \gamma^\nu) \cdot B = \gamma^\mu \cdot (\gamma^\nu \cdot B) \tag{7.43}$$

The expression (7.43) can be verified using

$$(\gamma^\mu \wedge \gamma^\nu) \cdot (\gamma_\alpha \wedge \gamma_\beta) = \delta^\mu_\beta \delta^\nu_\alpha - \delta^\mu_\alpha \delta^\nu_\beta, \tag{7.44}$$

which is a particular case of the identity found in the geometric algebra:

$$(a \wedge b) \cdot (c \wedge d) = (a \cdot d)(b \cdot c) - (a \cdot c)(b \cdot d) \tag{7.45}$$

If we make the inversion operation (denoted by ˜) on the generic multivector (7.37), we have

$$\tilde{M} = \alpha + a - B - bi + \beta i \tag{7.46}$$

that is, the bivector and the pseudovector change their sign.
(Remember, in fact, that for a vector of grade "r", we have $< \tilde{A} >_r = (-1)^{(r/2)(r-1)} A$).

Remember also that, due to the associative property of geometric product, i.e.,

$$a(bc) = (ab)c, \tag{7.47}$$

we have the associative rule

$$a \wedge (b \wedge c) = (a \wedge b) \wedge c \tag{7.48}$$

and the algebric identity

$$a(b \cdot c) + a \cdot (b \wedge c) = (a \cdot b)c + (a \wedge b) \cdot c. \tag{7.49}$$

We have also

$$a \cdot (b \wedge c) = (a \cdot b)c - (a \cdot c)b \tag{7.50}$$

(for the inner product between a vector and a bivector) and

$$a \wedge b \wedge c = -b \wedge a \wedge c = b \wedge c \wedge a \tag{7.51}$$

(for the outer product between a vector and a bivector).

More, in general, for the geometric product of a vector a and an s-vector B, one has

$$aB = a \cdot B + a \wedge B \tag{7.52}$$

where

$$a \cdot B = (1/2)(aB - (-1)^s Ba) \tag{7.53}$$

and

$$a \wedge B = (1/2)(aB + (-1)^s Ba) \tag{7.54}$$

Notice also that the pseudovectors are the *duals* of vectors: they are trivectors and are nothing else than the product of operator i (the pseudoscalar unit) by a vector. In other words, the pseudoscalar unit has the role of a multiplicative operator that determines the dual element of a multivector; in particular, the trivectors can be written as the duals of vectors and are also called pseudovectors. For instance,

$$\gamma_0 i = \gamma_1\gamma_2\gamma_3 \tag{7.55}$$

is a trivector, or the dual of the γ_0 vector.

Returning to the multivector (7.37), we can say that a multivector can always be decomposed into an even part M_+ [grade 0 (scalar), grade 2 (bivector), and grade 4 (pseudoscalar)] and an odd part M_- [grade 1 (vector) and grade 3 (trivector)], i.e.,

$$M_+ = \underset{(scalar)}{\alpha} + \underset{(bivector)}{B} + \underset{(pseudoscalar).}{\beta i} \tag{7.56}$$

$$M_- = \underset{(vector)}{a} + \underset{(pseudovector).}{bi} \tag{7.57}$$

The even multivectors or *spinors* (the sum of a scalar, a bivector, and a pseudoscalar) form a subalgebra (8-dimension), that is isomorphic to the Pauli algebra (see Chapter 5). More, in general, a multivector is said to be even if it is expressed as a sum of Clifford objects of even grade.

Going back to the Maxwell equations, we have seen that the electromagnetic field can be described by eight equations for six scalars (E_x, E_y, E_z, H_x, H_y, H_z) or two equations for an antisymmetric tensor ($F_{\mu\nu} = -F_{\nu\mu}$). Now, with the geometric algebra, we can describe the electromagnetic field by a single equation for a single multivector F, and this last possibility is simpler because multivector calculus has a more comprehensive geometric significance than the usual vector or tensor calculus.

We can see better this point with the consideration of the role of complex numbers in electrodynamics.

It is well known that the description of the electromagnetic waves is well done with complex numbers; for that reason the e.m. field is represented by complex quantities that are 'ad hoc': it is 'ad hoc' because only the real part of a complex quantity has physical significance. On the other hand, when the e.m. field is described by multivector calculus, complex quantities arise naturally with geometric and physical meanings.

It is, in fact, possible to write the Maxwell equations completely in terms of STA. Given the bivector of the e.m. field:

$$F = (1/2)F^{\mu\nu}\gamma_\mu \wedge \gamma_\nu \tag{7.58}$$

it is possible to substitute the two equations in tensorial form (7.19) and (7.20), by the two equations in multivector form (using also the rule (7.50) for the inner product and the rule (7.51) for the outer product):

$$\nabla \cdot F = (4\pi/c)j \tag{7.59}$$

$$\nabla \wedge F = 0 \tag{7.60}$$

where $\nabla = \gamma^\mu \partial_\mu$, which can be assembled in the unique equation

$$\nabla F = (4\pi/c)j \tag{7.61}$$

References

1. V. de Sabbata and M. Gasperini, *Introduction to Gravitation* (World Scientific, Singapore, 1985).
2. E. Cartan, *Compt. Rend.*, 174, 437, 593 (1922).
3. E. Cartan, *Ann. Ec. Norm.* 40, 325 (1923).
4. E. Cartan, *Ann. Sci. Ec. Norm.*, 41, 1 (1924).
5. E. Cartan, "Sur les variables à connexion affine et la théorie de la relativité généralisée", ed. Gauthier Villard, 1955.
6. F. W. Hehl, "Four Lectures on Poincaré Gauge Field Theory," in *Proceedings of the 6th Course of the International School of Cosmology and Gravitation on Spin, Torsion, Rotation, and Supergravity*, ed. P. G. Bergmann and V. de Sabbata (Plenum Press, New York, 1980) pp. 5–61.
7. M. Novello, *Phys. Lett.*, 59*A*, 105 (1976).
8. S. Hojman, M. Rosenbaum, and M. P. Ryan, *Phys. Rev.*, *D*17, 3141 (1978).
9. F. W. Hehl, P. von der Heyde, G. D. Kerlick, and J. M. Nester, *Rev. Mod. Phys.* 48, 393 (1976).
10. G. D. Kerlick, *Ann. Phys.*, 99, 127 (1976).
11. F. W. Hehl, P. von der Heyde, and G. D. Kerlick, *Phys. Rev.*, *D*10, 1066 (1974).
12. V. de Sabbata and M. Gasperini, *Nukleonika*, 25, *N*O. 11–12, 1980.
13. V. de Sabbata and M. Gasperini, *Nuovo Cimento Lett.*, 28, 181 (1980).
14. V. de Sabbata and M. Gasperini, *Nuovo Cimento Lett.*, 28, 229 (1980).
15. V. de Sabbata and M. Gasperini, *Phys. Lett.* 77*A*, 300 (1980).
16. V. de Sabbata and M. Gasperini, *Phys. Rev.* 23*D*, 2116 (1981).
17. V. de Sabbata and M. Gasperini, *Phys. Lett.* 83*A*, 115 (1981).
18. V. de Sabbata and M. Gasperini, *Nuovo Cimento Lett.* 30, 193 (1981).

8

Electromagnetic Field in Space and Time (Polarization of Electromagnetic Waves)

8.1 Electromagnetic (e.m.) Waves and Geometric Algebra

In order to see the description of e.m. waves in terms of geometric algebra, it is useful to express the bivector F of e.m. field in terms of the even subalgebra of the space-time algebra (STA) (formed by the scalar, bivector, and pseudoscalar). We remember that Equation (7.61) allows us to write F in terms of the current j (due to the fact that the ∇ operator, the same $\nabla = \gamma^{\mu}\partial_{\mu}$ that appears in the Dirac equation, is invertible; see, for instance, Reference [1], in which it is given that the Green function allows to solve F in terms of j). We have, in fact, from (7.60):

$$F = \nabla \wedge A, \tag{8.1}$$

where A is the vector potential of the e.m. field, which is invariant through a gauge transformation

$$A'(x) = A(x) + \nabla\alpha(x), \tag{8.2}$$

where $\alpha(x)$ is an arbitrary scalar function. Using the γ_0 frame, i.e., the laboratory system, we can write the bivector F in terms of the more familiar electric and magnetic fields E and B. In fact, Maxwell equations can be written as [2]:

$$(c^{-1}\partial_i + \nabla)F = 0, \tag{8.3}$$

where

$$F = E + iB, \tag{8.4}$$

where $E = E^k\sigma_k$ and $B = B^k\sigma_k$ are the *relative* vectors. Multiplying the bivector F by itself, we have the two Lorentz-invariant quantities (one scalar and one pseudoscalar):

$$F^2 = (|E|^2 - |B|^2) + i(E \cdot B) \tag{8.5}$$

In particular, for a plane e.m. wave that propagates in vacuum, we have $F^2 = 0$ and, then, F can be considered a light-type bivector. Notice that the expression (8.4) is actually more than a complex vector; in fact, "i" is more than a unit imaginary; it is the unit pseudoscalar that we have introduced (see Equation 7.35, where $i = \sigma_1\sigma_2\sigma_3 = \sigma_1 \wedge \sigma_2 \wedge \sigma_3$), and it appears in Equation 8.4 because the magnetic field is correctly described by the bivector $i\mathbf{B}$ and not by its dual $\mathbf{B}$.

It is easy to see that the relative vectors $E = E^k\sigma_k$ and $B = B^k\sigma_k$ are given explicity by

$$E = E^k\sigma_k = (1/2)(F - \gamma_0 F \gamma_0) \tag{8.6}$$

and

$$iB = iB^k\sigma_k = (1/2)(F + \gamma_0 F \gamma_0), \tag{8.7}$$

where

$$E^k = F^{k0},\ F^{ij} = -\varepsilon^{\mathrm{ijk}} B^k \qquad \text{(summation over } k\text{)}. \tag{8.8}$$

With the geometric significance of "i", the physical separation of F into electric and magnetic parts corresponds exactly to the geometric separation into vector and bivector parts. However, the usual form of vector calculus has preserved an artificial separation of electric and magnetic fields, though it was known that they compose a single physical entity. This has helped to disguise the significance of complex numbers in electrodynamics.

Electric and magnetic fields are in general represented as complex vectors without it being realized that the real part of one can be related to the imaginary part of the other so that only the single complex vector (8.4) is needed for a complete description of the field.

8.2 Polarization of Electromagnetic Waves

The compact description of the e.m. field (8.3) also yields a solution with more direct geometric significance as we will demonstrate in this section. The Maxwell equation admits a solution [3, 4]

$$F_+ = f \exp[i(\omega t - k \cdot r)] \tag{8.9}$$

that describes a monochromatic plane wave with frequency $\omega > 0$ and propagation vector $\mathbf{k}$. However, one may be surprised to learn that this wave is necessarily right circularly polarized, because that follows unequivocally only from the geometric meaning of multivector algebra.

To establish this property, we substitute Equation 8.9 in the Equation 8.3 using the fact that f, ω, and $\mathbf{k}$ are constant, so we have the Maxwell equation in the form

$$[(\omega/c) - k]F = 0. \tag{8.10}$$

Multiplying by $[(\omega/c) + k]$ we have

$$[(\omega^2/c^2) - k^2]F = 0. \quad (8.11)$$

Because F is not zero, this implies that $|z| = \omega/c$, as expected. So, Equation 8.10 can be written in the simpler form

$$\hat{k}F = F \quad (8.12)$$

where

$$\hat{k} = k/|k| = k/(\omega/c) \quad (8.13)$$

Then, substituting the Equation (8.4) in (8.12) we have that both even and odd parts of (8.12) give the same relation among the vector $\hat{k}$, E, and B, namely,

$$\hat{k}E = iB. \quad (8.14)$$

By using this equation (8.14) we can eliminate the magnetic field from Equation 8.4. In fact, remember that (see Equation 8.6 and Equation 8.7)

$$E = (1/2)(F - \gamma_0 F \gamma_0) \quad (8.6)$$

and

$$iB = (1/2)(F + \gamma_0 F \gamma_0) \quad (8.7)$$

where **E** represents the odd part of F and i**B** the even part:

$$F = \underset{\mathrm{E}}{(1/2)(F - \gamma_0 F \gamma_0)} + \underset{i\mathbf{B}}{(1/2)(F + \gamma_0 F \gamma_0)}. \quad (8.15)$$

We have

$$\begin{aligned}\hat{k}E &= (1/2)(\hat{k}F - \hat{k}\gamma_0 F \gamma_0) = (1/2)(F + \gamma_0 \hat{k} F \gamma_0) \\ &= (1/2)(F + \gamma_0 F \gamma_0) = iB,\end{aligned} \quad (8.16)$$

i.e., (8.14). It follows that we can eliminate the magnetic field from Equation (8.4) putting $\hat{k}E$ in the place of iB:

$$F = E + \hat{k}E = (1 + \hat{k})E = E(1 - \hat{k}) \quad (8.17)$$

The square of (8.12) gives $E^2 = B^2$, and we can write (8.14) in the form

$$\hat{\mathrm{E}}\hat{\mathrm{B}}\hat{\mathrm{k}} = i \quad (8.18)$$

The symbol ($\hat{}$) stands for unit vector.

We can say that E B k, in that order, compose a right-handed orthonormal frame of vectors.

To make the time dependence of the field explicit, note that from Equation (8.9) and Equation (8.17) we get

$$F(0,0) = f = E_0 + iB_0 = E_0(1 - \hat{k}). \tag{8.19}$$

Then, taking note of the fact that Equation 8.9 can also be written in this case as (see Appendix B)

$$\begin{aligned} F_+(r,t) = E(t) + iB(t) &= E_0(1-\hat{k})\exp[-i\hat{k}(\omega t - k\cdot r)] \\ &= f exp[-i\hat{k}(\omega t - k \cdot r)] \end{aligned} \tag{8.20}$$

so that at any point on the plane $k \cdot r = 0$ the field is

$$F = E(t) + iB(t) = E_0(1-\hat{k})\exp[-(i\hat{k}\omega t)]. \tag{8.21}$$

The vector part of this expression is

$$E(t) = E_0 \exp[-(i\hat{k}\omega t)] \tag{8.22}$$

$$= E_0(\cos\omega t - i\hat{k}\sin\omega t), \tag{8.23}$$

and, using (8.14), having $i\hat{k}E_0 = -B_0$, we have

$$E(t) = E_0\cos\omega t - B_0\sin\omega t \tag{8.24}$$

$$= E_0\cos\omega t + B_0\cos[(\pi/2) + \omega t], \tag{8.25}$$

and this shows explicitly that as t increases, the electric vector rotates clockwise in the plane as viewed by an observer facing the oncoming wave train, whereas the magnetic field vector follows 90° behind and the usual picture of a circularly polarized wave arises (see Figure 8.1) [5].

As regards the bivector part, we have

$$iB(t) = -E_0\hat{k}\exp[-(i\hat{k}\omega t)] \tag{8.26}$$

$$= -E_0\hat{k}(\cos\omega t - i\hat{k}\sin\omega t), \tag{8.27}$$

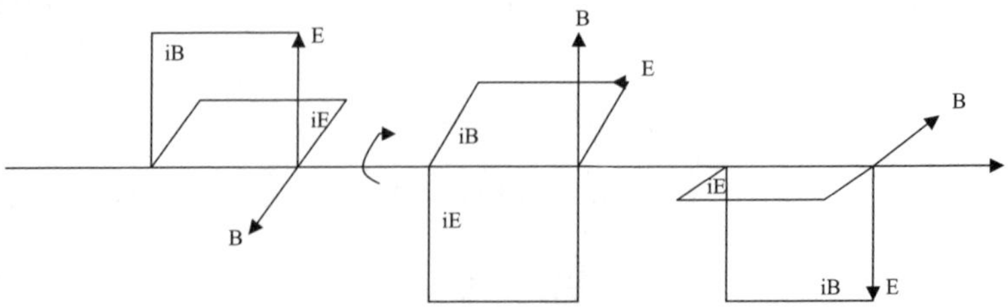

FIGURE 8.1
A graphic representation of a right circularly polarized electromagnetic wave as viewed by an observer facing the oncoming wave train.

that is, by using (8.14),

$$= iB_0 \cos \omega t + iE_0 \sin \omega t \quad (8.28)$$

$$= iB_0 \cos \omega t + iE_0 \cos[(3\pi/2) + \omega t], \quad (8.29)$$

which one can read also as

$$B(t) = B_0 \cos \omega t + E_0 \sin \omega t. \quad (8.30)$$

This shows that, as t increases, the magnetic bivector field $i\mathbf{B}$ also rotates clockwise along with the electric vector, whereas the electric bivector field follows 270° behind. Equation 8.25 and Equation 8.29 show that the electric vector field and the magnetic bivector field are in phase, whereas the magnetic vector field and the electric bivector field are in opposite phases (see Figure 8.1).

We note in passing that in the usual vector calculus the unit imaginary in the complex solution of the field equations generates rotation of **E** and **B** only [5].

8.3 Quaternion Form of Maxwell Equations from the Spinor Form of STA

In Chapter 7, Section 7.4, we have shown that Maxwell equations in STA can be assembled in the unique equation (see Eq. 7.61)

$$\nabla F = (4\pi/c)\, j, \quad (8.31)$$

where $\nabla = \gamma_\mu \partial_\mu$(summation over μ), and F is the electromagnetic bivector:

$$F = (1/2) F^{\mu\nu} \gamma_\mu / \gamma_\nu. \quad (8.32)$$

Equation 8.31 can be written as

$$[\gamma_0(\partial/\partial t) + \gamma_k(\partial/\partial x^k)]F = (4\pi/c)\, j \quad (8.33)$$

$$\text{(summation over } k; \quad k = 1, 2, 3)$$

Left multiplication (geometric) by γ_0 of both sides of the above equation gives

$$[\gamma_0\gamma_0(\partial/\partial t) + \gamma_0\gamma_k(\partial/\partial x^k)]F = (4\pi/c)\gamma_0\, j$$

or

$$[(\partial/\partial t) - \gamma_k\gamma_0(\partial/\partial x^k)]F = -(4\pi/c)(j \cdot \gamma_0 + j \wedge \gamma_0)$$

or

$$[(\partial/\partial t) - \sigma_k(\partial/\partial x^k)]F = -(4\pi/c)(c\rho + j), \quad (8.34)$$

where ρ is the charge density and j the current density vector. Here, the scalar part $j \cdot \gamma_0$ is the γ_0-time component of the vector j (i.e., the charge density), and the bivector $j \wedge \gamma_0$ is decomposed into the σ_k frame and shown to represent a spatial vector (i.e., the current density in 3-space) relative to an observer in the γ_0-frame (see also Chapter 5, Section 5.2, Equation 5.20).

In the γ_0-frame (i.e., in the laboratory system [5]) the electromagnetic field F can be expressed in the more familiar electric and magnetic fields E and B as (see Appendix A):

$$F = E + iB \tag{8.35}$$

where

$$E = F^{ko}\sigma_k = E^k\sigma_k \tag{8.36}$$
$$iB = iB^k\sigma_k, \tag{8.37}$$

B^k being given by

$$F^{ij} = -\epsilon^{\mathrm{ijk}}B^k \quad \text{(summation over } k\text{)}, \tag{8.38}$$

ϵ^{ijk} being the alternating tensor with $\epsilon^{123} = 1$.

By using Equation 8.36 and Equation 8.37, Equation 8.35 can be written as

$$F = (E^k + iB^k)\sigma_k. \tag{8.39}$$

By setting

$$\psi^k = E^k + iB^k, \tag{8.40}$$

Equation 8.39 can be written as

$$F = \psi^k\sigma_k = \psi. \tag{8.41}$$

By virtue of (8.41), the Maxwell Equation 8.34 assumes the form

$$[(\partial/\partial t) - \sigma_k(\partial/\partial x^k)]\psi^l\sigma_l = -(4\pi/c)(c\rho + j) \tag{8.42a}$$

or

$$[(\partial/\partial t) - \sigma_k(\partial/\partial x^k)]\sigma_l\psi^l = -(4\pi/c)(c\rho + j). \tag{8.42b}$$

Equation 8.42b can be written in expanded form:

$$\begin{aligned}&[(\partial/\partial t) - \sigma_k(\partial/\partial x^k)] \cdot (\sigma_l\psi^l) + [(\partial/\partial t) - \sigma_k(\partial/\partial x^k)] \wedge \sigma_l\tilde{\psi}^l \\ &\quad = -(4\pi/c)(c\rho + j).\end{aligned} \tag{8.43}$$

In order to maintain the exact correspondence between Hamilton's quaternions and Hestenes' spinors, it is necessary to replace ψ^i by $\tilde{\psi}^l$ in the above outer product.

Equation 8.43 can be separated into two independent equations:

$$[(\partial/\partial t) - \sigma_k(\partial/\partial x^k)] \cdot (\sigma_l \psi^l) = -4\pi\rho. \quad (8.44)$$

$$[(\partial/\partial t) - \sigma_k(\partial/\partial x^k)] \wedge \sigma_l \tilde{\psi}^l = -(4\pi/c)\, j. \quad (8.45)$$

This spinor form of Maxwell equations in geometric algebra corresponds to the quaternion form that was originally considered by Maxwell [6]. In fact, Maxwell obtained spinor equations by using Hamilton's quaternions.

8.4 Maxwell Equations in Vector Algebra from the Quaternion (Spinor) Formalism

We consider the spinor (quaternion) form of Maxwell Equation 8.44 and Equation 8.45 of the previous section (with units adopted such that $c = G = \hbar = 1$):

$$[(\partial/\partial t) - \sigma_k(\partial/\partial x^k)] \cdot (E + iB) = -4\pi\rho \quad (8.46)$$

and

$$[(\partial/\partial t) - \sigma_k(\partial/\partial x^k)] \wedge (E - iB) = -4\pi j, \quad (8.47)$$

where

$$E + iB = (\sigma_l \psi^l) = \psi \quad (8.48)$$

$$E - iB = (\sigma_l \tilde{\psi}^l) = \tilde{\psi}. \quad (8.49)$$

Equation 8.46 gives

$$-\sigma_k(\partial/\partial x^k) \cdot E - \sigma_k(\partial/\partial x^k) \cdot iB = -4\pi\rho,$$

that is,

$$\mathrm{divE} + \mathrm{idivB} = 4\pi\rho, \quad (8.50)$$

from which we get

$$\mathrm{divE} = 4\pi\rho \quad (\mathrm{M}\,4)$$

$$\mathrm{divB} = 0 \quad (\mathrm{M}\,2)$$

Next, Equation 8.47 gives

$$(\partial E/\partial t) - i(\partial B/\partial t) - \sigma_k(\partial/\partial x^k) \wedge E + i\sigma_k(\partial/\partial x^k) \wedge B = -4\pi j$$

or

$$(\partial E/\partial t) - i(\partial B/\partial t) - i\sigma_k(\partial/\partial x^k) \times E + ii\sigma_k(\partial/\partial x^k) \times B = -4\pi j$$

(using the formula $a \wedge b = ia \times b$)

or

$$(\partial E/\partial t) - i(\partial B/\partial t) - i\sigma_k(\text{rotE})_k - \sigma_k(\text{rotB})_k = -4\pi j$$

or

$$(\partial E/\partial t) - i(\partial B/\partial t) - i\,\text{rotE} - \text{rotB} = -4\pi j.$$

Finally, changing signs on both sides, we get

$$(\text{rotB} - \partial E/\partial t) + i(\text{rotE} + \partial B/\partial t) = 4\pi j,$$

from which we get

$$\text{rotB} - (\partial E/\partial t) = 4\pi j \tag{M 3}$$

$$\text{rotE} + (\partial B/\partial t) = 0 \tag{M 1}$$

Thus, we obtain the standard form of Maxwell equations describing electromagnetic field in the usual vector algebra

$$\text{rotE} + (\partial B/\partial t) = 0 \tag{M 1}$$

$$\text{divB} = 0 \tag{M 2}$$

$$\text{rotB} - (\partial E/\partial t) = 4\pi j \tag{M 3}$$

$$\text{divE} = 4\pi\rho \tag{M 4}$$

The subtlety of the spinor (quaternion) form for developing electrodynamics with integral spin is lost in the above formulation in vector algebra.

8.5 Majorana–Weyl Equations from the Quaternion (Spinor) Formalism of Maxwell Equations

In 1928, H. Weyl [7] represented the Dirac equation by a pair of equations

$$[(\partial/\partial t) + \bar{\sigma}(\partial/\partial \bar{x})]\psi_R + im\psi_L = 0 \tag{8.51a}$$

$$[(\partial/\partial t) - \bar{\sigma}(\partial/\partial \bar{x})]\psi_L + im\psi_R = 0 \tag{8.51b}$$

where $\sigma_k(k = 1, 2, 3)$ are the Pauli matrices, and Ψ_R, $\psi_L$are 2-component spinors. In the limit $m \Longrightarrow 0$, Equation (8.51a, b) transform into the Weyl equations [8]

$$[(\partial/\partial t) + \bar{\sigma}(\partial/\partial \bar{x})]\psi_R = 0 \tag{8.52a}$$

$$[(\partial/\partial t) - \bar{\sigma}(\partial/\partial \bar{x})]\psi_L = 0. \tag{8.52b}$$

These equations are similar in nature to the Maxwell equations in quaternion form.

Ettore Majorana first noted the quantum mechanical nature of the Maxwell equations. This is evident from his unpublished manuscript [9, 10] (kept in Domus Galilaeana in Pisa), in which he clarified the reason for the similarity in nature of the Maxwell and Weyl equations by introducing the transformation law for the vectors $\bar{\psi}_R$ and $\bar{\psi}_L$ (one should not get confused with spinors ψ_R, ψ_L and vectors $\bar{\psi}_R$, $\bar{\psi}_L$ used by Majorana; vectors are made distinct from spinors by putting a "bar" overhead) under a rotation of infinitesimal angle $\bar{\varphi}$

$$\bar{\psi}'_{R,L} = \bar{\psi}_{R,L} - [\bar{\varphi}, \bar{\psi}_{R,L}]. \tag{8.53}$$

Dropping the labels R and L, we can write this as

$$\begin{aligned} \psi'_k &= \psi_k - \epsilon_{kil} \quad \varphi_i \psi_l \\ &= \psi_k + \epsilon_{ikl} \quad \varphi_i \psi_l, \end{aligned} \tag{8.54}$$

where ϵ_{ikl} is the alternating tensor with $\epsilon_{123} = 1$. Equation 8.54 can be put in the form

$$\psi'_k = \psi_k + i \quad (S_i)_{kl} \quad \varphi_i \psi_l \tag{8.55}$$

by setting

$$(S_i)_{kl} = -i\epsilon_{ikl}. \tag{8.56}$$

The quantities $(S_i)_{kl}$, defined by (8.56), are spin angular momentum operators. By using them, one can represent Maxwell equations as given by Majorana:

$$[(\partial/\partial t) + \bar{S}(\partial/\partial \bar{x})]\psi_R = 0, \quad \text{div } \bar{\psi}_R = 0, \tag{8.57a}$$

$$[(\partial/\partial t) - \bar{S}(\partial/\partial \bar{x})]\psi_L = 0. \quad \text{div } \bar{\psi}_L = 0. \tag{8.57b}$$

These are of the same form as the Weyl equations supplemented by the equations div $\bar{\psi}_R =$ div $\bar{\psi}_L = 0$.

Now, we derive the spinor form of Maxwell equations in geometric algebra that corresponds to the Weyl form of Maxwell equations. For this purpose we consider a rigid body rotating about a reference point O with the paths of all points of it parallel to some plane. We can represent this plane by a bivector of unit modulus 'in', where n is the unit vector that specifies the rotation axis ON (see Figure 5.2). If a vector a in the Euclidean space E_3 is rotated through an infinitesimal angle ϑ about the n-direction to the vector a', then we have

$$a' = Ra\tilde{R}, \tag{8.58}$$

where R is a spinor given by (see Equation 5.60 and Equation 5.61)

$$R = \exp(\text{in}\vartheta/2). \tag{8.59}$$

For infinitesimal rotation, retaining terms up to the first order in ϑ, we have from Equation 8.58 and Equation 8.59 (see Equation 5.63)

$$a' = Ra\tilde{R} = a + i(n\vartheta \wedge a) \tag{8.60}$$

that is,

$$a' = a - n\vartheta \times a, \tag{8.61}$$

that is,

$$a_k' = a_k - \epsilon_{\mathrm{kil}}\vartheta_i a_l, \tag{8.62}$$

where ϵ_{kil} is the alternating tensor with $\epsilon_{123} = 1$, and ϑ_i are the covariant components of the vector $n\vartheta$ (n is the unit vector).

Now, we set

$$(S_i)_{kl} = -i\epsilon_{\mathrm{ikl}} = i\epsilon_{\mathrm{kil}}, \tag{8.63}$$

that is,

$$i(S_i)_{kl} = -\epsilon_{\mathrm{kil}}. \tag{8.64}$$

By virtue of (8.64) we can write (8.62) as

$$a_k' = a_k + i(S_i)_{kl}\vartheta_i a_l. \tag{8.65}$$

We note that the above transformation law is identical with the transformation law (8.55) for the vector $\bar{\psi}_R$ and $\bar{\psi}_L$ under rotation of an infinitesimal angle $\bar{\varphi}$. So, $(S_i)_{kl}$ defined by (8.63) are interpreted as spin angular momentum operators in this case.

In Section 8.3 we noted that in vacuum the spinor form of Maxwell equations are (see Equation 8.44 and Equation 8.45)

$$[(\partial/\partial t) - \sigma_k(\partial/\partial x^k)] \cdot \psi = 0 \tag{8.66a}$$

$$[(\partial/\partial t) - \sigma_k(\partial/\partial x^k)] \wedge \tilde{\psi} = 0 \tag{8.66b}$$

where

$$\psi = \sigma_l \psi^l = E + iH \tag{8.67a}$$

$$\tilde{\psi} = \sigma_l \tilde{\psi}^l = E - iH \tag{8.67b}$$

By using the operators $(S_i)_{kl}$ we can write Equation (8.66a, b) in the form

$$[(\partial/\partial t) - (S_i)_{kl}(\partial/\partial x^k)] \cdot \psi = 0 \tag{8.68a}$$

$$[(\partial/\partial t) - (S_i)_{kl}(\partial/\partial x^k)] \wedge \tilde{\psi} = 0 \tag{8.68b}$$

This spinor form of Maxwell equations in geometric algebra corresponds to the Majorana–Weyl form of Maxwell equations. We note in passing that unlike Equations 8.57a, b, the equations just assumed at are not to be supplemented by the divergence relations separately, as they are contained in Equation 8.68a.

Note that if we compare the Weyl Equations 8.52a, b and Maxwell equations as given by Majorana (8.57a, b) or (8.68a, b), they are found to be similar in nature. Now, Weyl equations refer to the particle of half spin, whereas Maxwell equations refer to particle of integral spin. This seems to give some hint of the problem of supersymmetry.

Appendix A: Complex Numbers in Electrodynamics

Electromagnetic Field: Transition from STA to Pauli Algebra

As the relation

$$\sigma_k = \gamma_k \gamma_0, \quad (k = 1, 2, 3) \tag{A.1}$$

satisfies

$$(1/2)(\sigma_i \sigma_j + \sigma_j \sigma_i) = -(1/2)(\gamma_i \gamma_j + \gamma_j \gamma_i) = \delta_{ij}, \tag{A.2}$$

the three space-time bivectors $\{\sigma_k\}$ generate the Pauli algebra spanned by the eight independent basis elements [5, 11, 12, 13]

$$1, \{\sigma_k\}, \{i\sigma_k\}, i, \tag{A.3}$$

which is an even subalgebra of STA. Because of the relation (A.1), the Pauli algebra (A.3) is identified with the algebra for the three-dimensional Euclidean space relative to the timelike vector γ_0. It is evident from the above fact that the separation of the six space-time bivector $\{\sigma_k, i\sigma_k\}$ into relative vectors $\{\sigma_k\}$ and relative bivectors $\{i\sigma_k\}$ of three-dimensional Euclidean space is a frame-dependent operation. The operation enables one to translate relativistic quantities into observables in a given three-dimensional frame. In what follows, we find the separation of the space-time bivector F representing the electromagnetic field into relative vectors and relative bivectors.

The electromagnetic field in space-time is given by the bivector

$$F = (1/2)F^{\mu\nu}\gamma_\mu \wedge \gamma_\nu, \qquad F^{\mu\nu} = -F^{\nu\mu}. \tag{A.4}$$

Expanding F in (A.4), we may write

$$F = \left[F^{10}\gamma_1\gamma_0 + F^{20}\gamma_2\gamma_0 + F^{30}\gamma_3\gamma_0\right] + \left[F^{32}\gamma_3\gamma_2 + F^{13}\gamma_1\gamma_3 + F^{21}\gamma_2\gamma_1\right] \tag{A.5}$$

The γ_0 vector maps F into $F\gamma_0$, which can be separated into antisymmetric and symmetric parts as

$$F\gamma_0 = F \cdot \gamma_0 + F \wedge \gamma_0, \tag{A.6}$$

where

$$F \cdot \gamma_0 = (1/2)(F\gamma_0 - \gamma_0 F) \tag{A.7a}$$

$$F \wedge \gamma_0 = (1/2)(F\gamma_0 + \gamma_0 F) \tag{A.7b}$$

In view of (A.5), one may express (A.7a) and (A.7b) as

$$F \cdot \gamma_0 = (1/2)(F\gamma_0 - \gamma_0 F) = F^{10}\gamma_1 + F^{20}\gamma_2 + F^{30}\gamma_3, \tag{A.8a}$$

$$F \wedge \gamma_0 = (1/2)(F\gamma_0 + \gamma_0 F) = i\left[F^{32}\gamma_1 + F^{13}\gamma_2 + F^{21}\gamma_3\right]. \tag{A.8b}$$

Now, the relative vector (electric) and relative bivector (magnetic field) are given, respectively, by

$$(1/2)(F - \gamma_0 F \gamma_0) = E^k \sigma_k = E, \tag{A.9a}$$

where

$$E^k = F^{k0} \tag{A.9b}$$

and

$$(1/2)(F + \gamma_0 F \gamma_0) = iB^k \sigma_k = iB, \tag{A.10a}$$

where

$$F^{ij} = -\epsilon^{\mathrm{ijk}} B^k \quad \text{(summation over } k\text{)}, \tag{A.10b}$$

ϵ^{ijk} being the alternating tensor with $\epsilon^{123} = 1$.

Thus, the separation of the space-time bivector F representing the electromagnetic field into the γ_0-system is given by

$$F = (F\gamma_0)\gamma_0 = E + iB, \tag{A.11}$$

where the electric vector E and the magnetic vector B are both spatial vectors as given, respectively, by (A.9a) and (A.10a), and iB is a spatial bivector representing the magnetic field (see Reference [5]).

Of course, one may write the result (A.11) directly from (A.5) without having the important relations

$$E = (1/2)(F - \gamma_0 F \gamma_0), \tag{A.12}$$

$$iB = (1/2)(F + \gamma_0 F \gamma_0). \tag{A.13}$$

From the foregoing we have

$$E\gamma_0 = -\gamma_0 E, \quad iB\gamma_0 = \gamma_0(iB), \tag{A.14}$$

i.e., γ_0 anticommutes with E but commutes with iB. Also, the space conjugation of the electric vector E and magnetic bivector iB exhibits that

$$\gamma_0 E \gamma_0 = -E, \tag{A.15}$$

$$\gamma_0(iB)\gamma_0 = iB, \tag{A.16}$$

which show that the electric vector changes sign whereas the magnetic bivector remains invariant under space conjugation.

Appendix B: Plane-Wave Solutions to Maxwell Equations — Polarization of e.m. Waves

Compare the geometric products of $(E_0 + \hat{k}E_0)$ with $\exp[-i\hat{k}(\omega t - k \cdot x)]$ and with $\exp[i(\omega t - k \cdot x)]$, where the former has only scalar and bivector parts and the latter, which has only scalar and pseudoscalar parts, may be treated as "formally complex."

First, consider the product $(E_0 + \hat{k}E_0)\exp[-i\hat{k}(\omega t - k \cdot x)]$:

$$E_0 \exp[-i\hat{k}(\omega t - k \cdot x)] = \text{vector part only.}$$
$$\hat{k}E_0 \exp[-i\hat{k}(\omega t - k \cdot x)] = \text{bivector part only.}$$

Summing up together, we have

$$\begin{aligned} &(E_0 + \hat{k}E_0)\exp[-i\hat{k}(\omega t - k \cdot x)] \\ &= [E_0 \cos(\omega t - k \cdot x) - B_0 \sin(\omega t - k \cdot x)] \quad \text{(vector part)} \\ &+[iB_0 \cos(\omega t - k \cdot x) + iE_0 \sin(\omega t - k \cdot x)] \quad \text{(bivector part).} \end{aligned} \tag{B.1}$$

Next, consider the product $(E_0 + \hat{k}E_0)\exp[i(\omega t - k \cdot x)]$:

$$E_0 \exp[i(\omega t - k \cdot x)] = \text{vector part + bivector part}$$
$$\hat{k}E_0 \exp[i(\omega t - k \cdot x)] = \text{vector part + bivector part}$$

Summing up together, we get

$$\begin{aligned} &(E_0 + \hat{k}E_0)\exp[i(\omega t - k \cdot x)] \\ &= [E_0 \cos(\omega t - k \cdot x) - B_0 \sin(\omega t - k \cdot x)] \quad \text{(vector part)} \\ &+[iB_0 \cos(\omega t - k \cdot x) + iE_0 \sin(\omega t - k \cdot x)] \quad \text{(bivector part).} \end{aligned} \tag{B.2}$$

Equations B.1 and B.2 show that both the products give the same result. This is because, in this case, the Maxwell equation in multivector algebra assumes the form

$$\hat{k}F = F, \tag{B.3}$$

which in turn yields

$$\hat{k}E = iB. \tag{B.4}$$

So, for our consideration, we may take any one of the above products. Thus, for a simple circularly polarized wave $F_+(x, t)$ with constant frequency $\omega > 0$

and propagating vector k, we can write

$$\begin{aligned} F_+(x,t) &= f\exp[i(\omega t - k\cdot x)] \\ &= f\exp[i\omega(t - k\cdot x/\omega)] \\ &= f\exp[i\omega(t - \hat{k}\cdot x/c)] \qquad \{\text{being } k = \hat{k}|k| = \hat{k}(\omega/c)\} \\ &= (1+\hat{k})E_0\exp(i\omega s) \\ &= (1+\hat{k})E_0 z(s), \end{aligned} \tag{B.5}$$

where

$$f = (1+\hat{k})E_0 \tag{B.5a}$$

$$s = t - \hat{k}\cdot x/c \tag{B.5b}$$

$$z(s) = \exp(i\omega s). \tag{B.5c}$$

Similarly, by changing the orientation of the "generator" in (B.5), we can express the left circularly polarized wave $F_-(x,t)$ as

$$F_-(x,t) = (1+\hat{k})E_0 z(s) \tag{B.6}$$

where

$$z(s) = \exp(-i\omega s) \tag{B.6a}$$

In both the solutions (B.5) and (B.6), the frequency is considered to be positive. In the general case, we consider the frequency to take both positive and negative values by associating its sign with the polarization of the wave, $F_+(x,t)$ and $F_-(x,t)$, being, respectively, positive and negative frequency solutions.

For a wave packet propagating in the direction of $\hat{k}$, one may write $F(x,t)$ in the general form

$$F(x,t) = f z(s) \tag{B.7}$$

where

$$f = (1+\hat{k})e \tag{B.8}$$

$$s = t - \hat{k}\cdot x/c, \tag{B.9}$$

e being a constant unit vector orthogonal to $\hat{k}$. Function $z(s)$ is the Fourier transform of the function $\alpha(\omega)$ that satisfies the Dirichlet's conditions, i.e.,

$$z(s) = \int_{-\infty}^{\infty} \alpha(\omega)e^{i\omega s}d\omega, \tag{B.10}$$

where $\exp(i\omega s)$ is the kernel of the transform. Equation B.10 can be expressed as

$$z(s) = \int_{-\infty}^{\infty} [\alpha_+(\omega)e^{i\omega s} + \alpha_-(\omega)e^{-i\omega s}]d\omega \tag{B.11}$$

where

$$\alpha_{\pm}(\omega) = \alpha(\pm\omega). \tag{B.12}$$

The functions $\alpha_+(\omega)$ and $\alpha_-(\omega)$ are the components of the wave packet and describe, respectively, the right and left circular polarization. The overall phase of $\alpha(\omega)$ depends on the selection of the constant unit vector e in the plane orthogonal to the propagating vector k. Equation B.7 and Equation B.11 exhibit that the "formally complex" function $z(s)$ describes the properties of the wave packet.

From Equation B.8 we have

$$|f|^2 = 2. \tag{B.13}$$

So, the energy-density of the field is given by

$$|z|^2 = (1/2)|F|^2 = (1/2)(E^2 + B^2). \tag{B.14}$$

References

1. S. Gull, A. Lasenby, and C.Doran, *Found. Phys.*, 23, 1329, (1993).
2. D. Hestenes, *Space Time Algebra* (Gordon and Breach, New York, 1996).
3. D. Hestenes, *Am. J. Phys.*, 39, 1013 (1971).
4. T. G. Vold, *Am. J. Phys.*, 61, 505 (1993).
5. B. K. Datta and V. de Sabbata, "Geometric algebra and polarization of electromagnetic Waves" in *Tensor*, ed. Tomoaki Kawaguchi (Tensor Society, Chigasaki, Japan), N.S. 61, 181–190 (1999).
6. J. C. Maxwell, *A Treatise on Electricity and Magnetism* (Oxford University Press, Oxford, 1873), Volume 5 , page 2.
7. H. Weyl, *Gruppentheory und Quantummechanik*, (Verlag von S. Hirzel, Leipzig, 1928).
8. H. Weyl, "Elektron und Gravitation," *Z. Phys.*, 56, 330–572 (1929).
9. R. Mignani, E. Recami, and M. Baldo, "About a Dirac-like equation for the photon according to Ettore Majorana," *Nuovo Cimento Lett.*, 11, 568 (1974).
10. Yu. P. Stepanovsky, "Ettore Majorana and Matvei Bronstein (1906–1938): Men and Scientists," in *Advances in the Interplay between Quantum and Gravity Physics*, ed. P. G. Bergmann and V. de Sabbata (Kluwer Academic Publishers, Dordrecht, The Netherlands, 2002), pp. 435–458.
11. B. K. Datta, "Physical theories in space-time algebra," in *Quantum Gravity*, ed. P. G. Bergmann, V. de Sabbata, and H. -J. Treder (World Scientific, Singapore 1996), pp. 54–79.
12. B. K. Datta, R. Datta, and V. de Sabbata, *Found. Phys. Lett.*, 11, 83 (1998).
13. B. K. Datta, V. de Sabbata, and L. Ronchetti, *Nuovo Cimento*, 113 B, 711 (1998).

9

General Observations and Generators of Rotations (Neutron Interferometer Experiment)

9.1 Review of Space-Time Algebra (STA)

We give a brief resumé of STA as developed by Hestenes [1,2,3]. STA is the Clifford algebra of real four-dimensional space-time with a thoroughgoing geometric interpretation. It is built out of objects with direct geometric interpretations; the properties of these objects are specified by introducing algebraic operations that directly determine their interrelations. Furthermore, STA derives its potency from the fact that both the elements and the operations of the algebra are endowed with direct geometric intepretation.

The geometric product of a generic proper vector a with itself is a scalar quantity describing the metric of space-time:

$$a^2 > 0 \text{ iff a is a timelike vector,} \tag{9.1a}$$

$$a^2 = 0 \text{ iff a is a lightlike vector,} \tag{9.1b}$$

$$a^2 < 0 \text{ iff a is a space-like vector.} \tag{9.1c}$$

The geometric product ab of two proper vectors a and b can be understood geometrically by separating it into symmetric and antisymmetric parts:

$$ab = a \cdot b + a \wedge b, \tag{9.2a}$$

where

$$a \cdot b \equiv (1/2)(ab + ba) = b \cdot a, \tag{9.2b}$$

$$a \wedge b \equiv (1/2)(ab - ba) = -b \wedge a. \tag{9.2c}$$

Equation 9.1 tells us that $a \cdot b$ is a scalar quantity, the usual inner product of space-time vectors. Here 'scalar' means 'real number'.

The quantity $a \wedge b$, called the "outer product" of a, b, is a (proper) bivector or 2-vector. The geometric product obeys the associative rule.

The inner and outer product of a vector a and a bivector B are defined respectively, by

$$a \cdot B \equiv (1/2)(aB - Ba) = -B \cdot a, \tag{9.3a}$$

$$a \wedge B \equiv (1/2)(aB + Ba) = B \wedge a. \tag{9.3b}$$

Then, we have

$$aB = a \cdot B + a \wedge B. \tag{9.3c}$$

Equation (9.3a) and Equation (9.2b, c) give

$$a \cdot (b \wedge c) = a \cdot bc - a \cdot cb = -(b \wedge c) \cdot a. \tag{9.4}$$

9.1.1 Note

1. $a \cdot B$ is a vector.
2. $a \wedge B$ is a trivector or 3-vector.
3. Every trivector in STA can be factored into an outer product of three vectors.

The inner and outer product of a vector a and a trivector T are defined, respectively, by

$$a \cdot T \equiv (1/2)(aT + Ta) = T \cdot a, \tag{9.5a}$$

and

$$a \wedge T \equiv (1/2)(aT - Ta) = -T \wedge a. \tag{9.5b}$$

Thus

$$aT = a \cdot T + a \wedge T. \tag{9.5c}$$

Equation (9.5a), Equation (9.3c) and Equation 9.2 give

$$\begin{aligned} a \cdot (b \wedge B) = a \cdot (bB - b \cdot B) &= a \cdot bB - a \cdot (b \cdot B) \\ &= a \cdot bB - a \wedge b \cdot B, \end{aligned} \tag{9.6}$$

where a, b are vectors and B is a bivector. $a \wedge T$ is a 4-vector or pseudoscalar. Unit pseudoscalar, denoted by i, assigns an orientation to space-time. Every pseudoscalar is a scalar multiple of i. It can be shown that

$$i^2 = -1, \tag{9.7a}$$

$$ai = -ia. \tag{9.7b}$$

From the above it follows that

$$a \wedge i \equiv (1/2)(ai + ia) = 0, \tag{9.8a}$$

$$a \cdot i \equiv (1/2)(ai - ia) = ai. \tag{9.8b}$$

ai is a trivector, called the dual of a. Every trivector T is the dual of some vector t:

$$T = ti. \tag{9.9a}$$

Multiplying on the right by i and using (9.7a), we get

$$Ti = -t. \tag{9.9b}$$

So, the dual of a trivector T is a unique vector. This establishes an isomorphism of the linear space of all trivectors to the space of all vectors. So, trivectors are often called pseudovectors.

9.1.2 Multivectors

A generic element of STA is called a (proper) multivector. Every proper multivector M is a sum of Clifford objects of arbitrary grade (grade zero = scalar, grade 1 = vector, grade 2 = bivector or 2-vector, grade 3 = trivector or 3-vector, grade 4 = 4-vector or pseudoscalar). Thus, M can be expressed as

$$M = < M >_0 + < M >_1 + < M >_2 + < M >_3 + < M >_4, \tag{9.10}$$

where $< M >_k$ is a Clifford object of grade k or denotes the k-vector part of M.

A multivector M in STA is said to be even if

$$< M >_1 = < M >_3 = 0.$$

The even multivectors constitute a subalgebra of the full STA.

9.1.3 Reversion

The reverse of M, denoted by $\tilde{M}$, is defined by

$$\tilde{M} = < M >_0 + < M >_1 - < M >_2 - < M >_3 + < M >_4 \tag{9.11}$$

The reverse of a product equals the product of the reverse:

$$(AB)^{\sim} = \tilde{B}\tilde{A} \tag{9.12}$$

9.1.4 Lorentz Rotation $\mathbb{R}$

Any Lorentz rotation $\mathbb{R}$, which maps a generic proper vector a into the vector a', can be written in the canonical form

$$a \Longrightarrow a' = \mathbb{R}(a) = Ra\tilde{R}. \tag{9.13a}$$

Here, R is an even multivector, unique except for sign, with the property

$$R\tilde{R} = 1. \tag{9.13b}$$

R is called a spinor.

9.1.5 Two Special Classes of Lorentz Rotations: Boosts and Spatial Rotations

A Lorentz rotation $\mathcal{L}(a) = La\tilde{L}$, which takes a unit timelike vector u into the vector v, is said to be a boost of u into v if it leaves the vectors orthogonal to the $v \wedge u$-plane invariant.

Any vector a can be expressed as the sum

$$a = a_{\|} + a_{\perp} \tag{9.14a}$$

with the component $a_{\|}$ in the $v \wedge u$-plane given by

$$a_{\|} = a \cdot (v \wedge u)(v \wedge u)^{-1}, \tag{9.14b}$$

$$a_{\perp} = a \wedge (v \wedge u)(v \wedge u)^{-1}, \tag{9.14c}$$

A Lorentz rotation $\mathbb{U}(a) = Ua\tilde{U}$ is said to be spatial rotation if it leaves a timelike vector u invariant:

$$Uu\tilde{U} = u \tag{9.15a}$$

or equivalently,

$$UU^{\dagger} = 1, \qquad U^{\dagger} \equiv u\tilde{U}u. \tag{9.15b}$$

The set of all Lorentz rotations satisfying (9.15) is the group of spatial rotations in the spacelike hypersurface with normal u, called the little group of u.

Any Lorentz rotation can be uniquely expressed as a spatial rotation followed by a boost of a given timelike vector u:

$$R = LU. \tag{9.16}$$

9.1.6 Magnitude

The magnitude (or modulus) of a multivector M is determined by

$$|M| = [< M\tilde{M} >_o]^{1/2} \tag{9.17}$$

The inverse M^{-1} of a multivector M is defined by

$$MM^{-1} = 1 \tag{9.18}$$

If $M\tilde{M} = |M|^2$, the inverse of M exists and can be expressed as

$$M^{-1} = \tilde{M}/|M|^2 \tag{9.19}$$

It follows that every vector has an inverse, so it is possible to divide by vectors. The vector division is made possible by (9.2a, b, c). So, the STA is an associative division algebra.

9.1.7 The Algebra of a Euclidean Plane

The Clifford algebra for two-dimensional Euclidean space is generated by two orthonormal vectors $\{\sigma_k\}$, $(k = 1, 2)$, and is spanned by

$$1, \quad \{\sigma_k\}, i, \tag{9.20}$$

where i is the unit pseudoscalar (highest grade multivector) for the space.

The basic properties of i are

$$i^2 = -1 \quad \Longrightarrow \quad i = \sqrt{-1}, \tag{9.21a}$$

$$\tilde{i} = -i, \tag{9.21b}$$

$$iM = Mi \tag{9.21c}$$

for every multivector M in the space, and

$$a \wedge b = \lambda i \tag{9.21d}$$

for any vector in the Euclidean plane. The scalar λ is positive if and only if the vectors make up a right-handed set in the order given.

Besides the property (9.21a) ascribed to the traditional unit imaginary, our i is a bivector. So, it has geometric and algebraic properties:

1. It is the unit of directed area.
2. It is also the generator of rotations in the plane.

In view of (9.20), any multivector M can be expressed as

$$M = a_0 + a_1\sigma_1 + a_2\sigma_2 + a_3 i \tag{9.22}$$

with scalar coefficients a_0, a_1, a_2, a_3. The algebra (9.20) is the geometric algebra of the i-plane and is denoted by $\mathcal{G}_2(i)$. Obviously, the $\mathcal{G}_2$ algebra is a four-dimensional linear space. We can write (9.22) as

$$M = < M >_+ + < M >_-, \tag{9.23}$$

where the even multivector part $< M >_+$ represents the plane spinor

$$< M >_+ = a_0 + i a_3 \tag{9.24}$$

and the odd multivector part $< M >_-$ represents the vector in the i-plane

$$< M >_- = a_1\sigma_1 + a_2\sigma_2. \tag{9.25}$$

So, $\mathcal{G}_2$ can be expressed as the sum of two linear spaces

$$\mathcal{G}_2 = \mathcal{G}_2^+ + \mathcal{G}_2^-, \tag{9.26}$$

where $\mathcal{G}_2^+$ is the two-dimensional linear space of spinors, and $\mathcal{G}_2^-$ is the two-dimensional vector space.. The algebra (9.24) is an even subalgebra $\mathcal{G}_2^+$ of

$\mathcal{G}_2$ or spinor algebra. Likewise, the algebra (9.25) is an odd subalgebra $\mathcal{G}_2^-$ of $\mathcal{G}_2$ or vector algebra. Every spinor in $\mathcal{G}_2^+$ represents a rotation–dilation in two-dimensional plane.

9.1.8 The Algebra of Euclidean 3-Space

The Clifford algebra for three-dimensional Euclidean space is generated by three orthonormal vectors $\{\sigma_k\}$ and is spanned by

$$1, \{\sigma_k\}, \{i\sigma_k\}, i, \tag{9.27}$$

where $i = \sigma_1\sigma_2\sigma_3$ is the unit pseudoscalar (highest grade multivector) for the space. The basic properties of i are

$$i^2 = -1, \tag{9.28a}$$

$$\tilde{i} = -i, \tag{9.28b}$$

$$iM = Mi \tag{9.28c}$$

for every multivector M in the space, and

$$a \wedge b \wedge c = \lambda i \tag{9.28d}$$

for every vector in the Euclidean 3-space. The scalar λ is positive if and only if the vectors make up a right-handed set in the order given. The unit i is the dextral unit pseudoscalar (trivector) giving the unit-oriented cube.

In view of (9.27), any multivector M of the space can be expressed as

$$M =< M >_0 + < M >_1 + < M >_2 + < M >_3 \tag{9.29}$$

with

$< M >_0 = \alpha,$	the scalar part of M,
$< M >_1 = a,$	the vector part of M,
$< M >_2 = ib,$	the bivector part of M expressed as a dual of a vector, and
$< M >_3 = i\beta,$	the pseudoscalar part of M.

So, (9.29) can be put in the form

$$M = \alpha + a + ib + i\beta \tag{9.30}$$

with scalar coefficient α and β. The algebra (9.27) is the geometric algebra of the i-space and is denoted by $\mathcal{G}_3$. It is also called the Pauli algebra, but in geometric algebra the three Pauli σ_k are no longer viewed as three matrix-valued components* of a single isospace vector but as three independent-basis vectors for real space.

Obviously, the $\mathcal{G}_3$ algebra or the Pauli algebra is an eight-dimensional linear space. We can write (9.30) as

$$M = < M >_+ + < M >_-, \tag{9.31}$$

where the even multivector part $< M >_+$ represents the spinor space

$$< M >_+ = \alpha + ib, \tag{9.32}$$

and the odd multivector part $< M >_-$ is given by

$$< M >_- = a + i\beta \tag{9.33}$$

So, $\mathcal{G}_3$ can be expressed as the sum of an even part $\mathcal{G}_3^+$ and an odd part $\mathcal{G}_3^-$:

$$\mathcal{G}_3 = \mathcal{G}_3^+ + \mathcal{G}_3^-. \tag{9.34}$$

$\mathcal{G}_3^+$ is closed under multiplication, so it is a subalgebra of $\mathcal{G}_3$, but $\mathcal{G}_3^-$ is not. $\mathcal{G}_3^+$ may be referred to as spinor algebra or even the subalgebra of $\mathcal{G}_3$(Pauli) algebra. Every spinor in $\mathcal{G}_3^+$ represents a rotation–dilation in three-dimensional space.

We have three linearly independent bivectors given by

$$i_1 = \sigma_2\sigma_3 = i\sigma_1, \quad i_2 = \sigma_3\sigma_1 = i\sigma_2, \quad i_3 = \sigma_1\sigma_2 = i\sigma_3. \tag{9.35}$$

So, any bivector B in $\mathcal{G}_3$ can be expressed as

$$B = \mathbb{R}_1 i_1 + \mathbb{R}_2 i_2 + \mathbb{R}_3 i_3 \tag{9.36}$$

with scalar coefficient $\mathbb{R}_k$. Thus, the set of all bivectors in $\mathcal{G}_3$ is a three-dimensional linear space with basis $\{i_1, i_2, i_3\}$. By using (9.35), the equation 9.36 can be expressed as

$$B = i(\mathbb{R}_1\sigma_1 + \mathbb{R}_2\sigma_2 + \mathbb{R}_3\sigma_3) = ib, \tag{9.37}$$

where b is a vector defined by

$$b = \mathbb{R}_1\sigma_1 + \mathbb{R}_2\sigma_2 + \mathbb{R}_3\sigma_3. \tag{9.38}$$

The bivector B is called the dual of the vector b. By using (9.36) and (9.37), one can express (9.32) as

$$< M >_+ = \alpha + \mathbb{R}_1 i_1 + \mathbb{R}_2 i_2 + \mathbb{R}_3 i_3, \tag{9.39}$$

---

* $\hat{\sigma}_1 = \begin{vmatrix} 0 & 1 \\ 1 & 0 \end{vmatrix}, \ \hat{\sigma}_2 = \begin{vmatrix} 0 & -j \\ j & 0 \end{vmatrix}, \ \hat{\sigma}_3 = \begin{vmatrix} 1 & 0 \\ 0 & -1 \end{vmatrix},$ where j is the unit imaginary

This shows that $\{1, i_1, i_2, i_3\}$ make up a basis for $\mathcal{G}_3^+$. Thus, $\mathcal{G}_3^+$ is a linear space of four dimensions. As the elements of $\mathcal{G}_3^+$ were called quaternions by Hamilton, $\mathcal{G}_3^+$ may also be referred to as quaternion algebra.

9.1.9 The Algebra of Space-Time

The Clifford algebra of real four-dimesional space-time is generated by four orthonormal vectors $\{\gamma_\mu\}$ and is spanned by

$$\begin{array}{ccccc} 1, & \{\gamma_\mu\}, & \{\sigma_k, i\sigma_k\}, & \{i\gamma_\mu\}, & i \\ \textit{scalar} & \textit{vectors} & \textit{bivectors} & \textit{pseudovectors} & \textit{pseudoscalar} \end{array} \tag{9.40}$$

$$(\mu = 0, 1, 2, 3; \qquad k = 1, 2, 3)$$

where i is the unit pseudoscalar (4-vector) for the space-time

$$i \equiv \gamma_0\gamma_1\gamma_2\gamma_3 = \sigma_1\sigma_2\sigma_3 = \gamma_5, \tag{9.41a}$$

$$\sigma_k \equiv \gamma_k\gamma_0. \tag{9.41b}$$

The even elements of the basis (9.40) for the space-time

$$1, \{\sigma_k, i\sigma_k\}, i \tag{9.42}$$

coincides with Pauli algebra because of the relation (9.41b). Thus, vectors in Pauli algebra become bivectors as viewed from (real) Dirac algebra. In STA the four Dirac γ_μ are no longer viewed as four matrix-valued components of a single isospace vector but as four independent basis vectors for real space-time. The pseudoscalar i anticommutes with the space-time vectors γ_μ.

In STA the reversion changes the sign of bivectors, leaving scalars and vectors unchanged. So, the reversion of an even multivector can be obtained by changing the sign of the bivectors.

The algebra (9.40) is the STA or real Dirac algebra, having 16 components, and is a 16-dimensional linear space. The algebra (9.42) is an even subalgebra of STA or real Dirac algebra with respect to the selection of the timelike vector γ_0, and is an eight-dimensional linear space of spinors.

9.2 The Dirac Equation without Complex Numbers

A Dirac spinor Ψ is represented by

$$\Psi = \begin{bmatrix} \psi_1 \\ \psi_2 \\ \psi_3 \\ \psi_4 \end{bmatrix} = \begin{bmatrix} \alpha_1 + j\beta_1 \\ \alpha_2 + j\beta_2 \\ \alpha_3 + j\beta_3 \\ \alpha_4 + j\beta_4 \end{bmatrix}, \tag{9.43}$$

where α's and β's are real numbers, and $j = \sqrt{-1}$ is the unit imaginary of the matrix algebra. The representation (9.43) in terms of the components

$\psi_1, \psi_2, \psi_3, \psi_4$ presumes a specific representation of the Dirac matrices:

$$\hat{\gamma}_0 = \begin{bmatrix} I & 0 \\ 0 & -I \end{bmatrix}, \quad \hat{\gamma}_0 = \begin{bmatrix} 0 & -\hat{\sigma}_k \\ \hat{\sigma}_k & 0 \end{bmatrix}, \tag{9.44a}$$

where I is a 2×2 unit matrix and $\hat{\sigma}_k$ are the Pauli matrices, which are traceless Hermitian matrices satisfying

$$\hat{\sigma_1}\hat{\sigma_2}\hat{\sigma_3} = jI. \tag{9.44b}$$

We introduce in the spinor space a basis [4]:

$$u_1 = \begin{bmatrix} 1 \\ 0 \\ 0 \\ 0 \end{bmatrix}, \quad u_2 = \begin{bmatrix} 0 \\ 1 \\ 0 \\ 0 \end{bmatrix}, \quad u_3 = \begin{bmatrix} 0 \\ 0 \\ 1 \\ 0 \end{bmatrix}, \quad u_4 = \begin{bmatrix} 0 \\ 0 \\ 0 \\ 1 \end{bmatrix}, \tag{9.45}$$

such that

$$\gamma_0 u_1 = u_1, \tag{9.46a}$$

$$i\sigma_3 u_1 = \gamma_2\gamma_1 u_1 = iu_1, \tag{9.46b}$$

$$u_2 = -i\sigma_2 u_1, \quad u_3 = \sigma_3 u_1, \quad u_4 = \sigma_1 u_1 \tag{9.46c}$$

Supposing (9.43) refers to this representation, one can write any Dirac spinor Ψ in the form

$$\Psi = \psi u_1 \tag{9.47}$$

where

$$\begin{aligned} \psi = \underset{(scalar)}{\alpha_1} &+ \underset{(bivectors)}{(\alpha_4\gamma_1\gamma_0 + \beta_4\gamma_2\gamma_0 + \alpha_3\gamma_3\gamma_0 + \beta_2\gamma_3\gamma_2 + \alpha_2\gamma_3\gamma_1 + \beta_1\gamma_2\gamma_1)} \\ &+ \underset{(pseudoscalar).}{\gamma_5\beta_3} \end{aligned} \tag{9.48}$$

The important thing is that the unit imaginary $j = \sqrt{-1}$ has been eliminated, and the ψ given by (9.48) is an even multivector that can be expressed as an element of STA by interpreting the γ's as vectors.

Dirac's equation for an electron with charge e and mass m in an external electromagnetic field is expressed by

$$(j\hbar\Box - (e/c)A)\Psi = mc\Psi, \tag{9.49}$$

where

$$\Box \equiv \gamma^\mu \partial_\mu, \quad \partial_\mu \equiv \partial/\partial x^\mu, \tag{9.50a}$$

and

$$A = A_\mu \gamma^\mu = A^\mu \gamma_\mu. \tag{9.50b}$$

Replacing $j = \sqrt{-1}$ by the bivector $\gamma_2\gamma_1$ and using (9.47), we can write (9.49) in the form

$$(\hbar\Box\psi\gamma_2\gamma_1 - (e/c)A\psi)\gamma_0 u_1 = mc\psi u_1, \tag{9.51}$$

with the help of (9.46a).

The coefficient of u_1 in (9.47) is an even multivector. (9.46c) implies that even a multivector operating on u_1 generates a complete basis for Dirac spinors; so, the coefficients of u_1 in (9.51) can be equated, although u_1 does not have an inverse. Therefore, Equation 9.51 yields

$$(\hbar\Box\psi\gamma_2\gamma_1 - (e/c)A\psi)\gamma_0 = mc\psi \tag{9.52a}$$

or, equivalently,

$$\hbar\Box\psi\gamma_2\gamma_1 - (e/c)A\psi = mc\psi\gamma_0. \tag{9.52b}$$

This is the Dirac equation in STA without unit imaginary $j = \sqrt{-1}$. By using Equation 9.47, a solution of one equation can be expressed as a solution of the other.

9.3 Observables and the Wave Function

In order to determine the geometrical significance of the wave function Ψ in Dirac's theory, it is required that certain bilinear functions of Ψ be tensors. The interpretation of these tensors as observables determines the physical significance of Ψ.

As $\tilde{i} = i$, $\tilde{\psi}$ can be obtained from ψ by changing the sign of its bivector part. As ψ is an even multivector, so is $\psi\tilde{\psi}$. However, $\psi\tilde{\psi}$ is invariant under reversion. So, its bivector part must vanish. Furthermore, because $i^2 = -1$, $\psi\tilde{\psi}$ can be put in the "polar form"

$$\psi\tilde{\psi} = \rho\exp(i\beta) = \rho\cos\beta + i\rho\sin\beta, \tag{9.53}$$

where ρ and β are scalars.

Also we know that

$$R\tilde{R} = 1. \tag{9.54a}$$

Thus, one can define R by the equation

$$R = [\rho\exp(i\beta)]^{-1/2}\psi \tag{9.54b}$$

or one may write

$$\psi = [\rho\exp(i\beta)]^{1/2}R. \tag{9.54c}$$

The quantity R in (9.54) determines a proper Lorentz transformation of a frame $\{\gamma_\mu\}$ into a frame $\{e_\mu\}$ according to the equation

$$e_\mu = R\gamma_\mu \tilde{R}. \tag{9.55}$$

$R = R(x)$ is a differentiable function of the space-time point x. Thus, Equation 9.55 specifies a differentiable set of four vector fields with values $e_\mu = e_\mu(x)$ at each point x determined by a proper Lorentz transformation of a fixed frame $\{\gamma_\mu\}$. This completely describes the geometrical significance of R. By virtue of Equation 9.55, the spinor R can be regarded as a representation of a Lorentz transformation.

Equation 9.54c and Equation 9.55 give

$$\psi\gamma_\mu\tilde{\psi} = \rho e_\mu. \tag{9.56}$$

Treating (9.56) as a generalization of (9.55), one may interpret the multiplication of e_μ by ρ as a dilation.

The spinor ψ represents a Lorentz transformation because, by Equation 9.56 it determines a rotation–dilation of the frame $\{\gamma_\mu\}$ into the frame $\{\rho e_\mu\}$.

It is to be noted that ψ does not operate in some abstract spin-space detached from space-time; it transforms space-time vectors into space-time vectors.

The physical interpretations of ρ and R are fixed by specifying the interpretations for the e_μ.

The quantity [4]

$$\rho v \equiv \rho e_0 = \psi\gamma_0\tilde{\psi} \tag{9.57a}$$

may be identified with the probability current of the Dirac theory. So, the timelike vector

$$v \equiv e_0 = R\gamma_0\tilde{R} \tag{9.57b}$$

can be interpreted as the (local) "world velocity" of a Dirac particle, whereas ρ is the probability density in the local rest frame determined by v.

The tensor components of the probability current with respect to the frame $\{\gamma_\mu\}$ are

$$\rho v^\mu = \rho v \cdot \gamma^\mu = (\psi\gamma_0\tilde{\psi}\gamma^\mu)_0. \tag{9.57c}$$

The conservation of probability is given by

$$\Box \cdot (\rho v) = \partial_\mu(\rho v^\mu) = 0. \tag{9.58}$$

The spacelike vector

$$s \equiv (\hbar/2)e_3 = (\hbar/2)R\gamma_3\tilde{R} \tag{9.59a}$$

can be identified as the (local) spin vector of the Dirac theory. The corresponding "current"

$$\rho s \equiv \rho(\hbar/2)e_3 = (\hbar/2)\psi\gamma_3\tilde{\psi} \tag{9.59b}$$

has components

$$\rho s_\mu = \rho s \cdot \gamma_\mu = (\hbar/2)(\psi\gamma_3\tilde{\psi}\gamma_\mu)_0. \tag{9.59c}$$

The "proper spin density" of the electron is ρS, where S is the (local) spin bivector:

$$S \equiv (\hbar/2)e_2e_1 = (\hbar/2)R\gamma_2\gamma_1\tilde{R} = (\hbar/2)Ri\sigma_3\tilde{R} = isv. \tag{9.60a}$$

The tensor components of S are given by

$$S^{\alpha\beta} = (S\gamma^\beta\gamma^\alpha)_0 = is \wedge v \wedge \gamma^\beta \wedge \gamma^\alpha = s_\mu v_\nu \epsilon^{\mu\nu\alpha\beta}, \tag{9.60b}$$

where the alternating tensor $\epsilon^{\mu\nu\alpha\beta}$ is defined by

$$\epsilon^{\mu\nu\alpha\beta} = -i\gamma^\mu \wedge \gamma^\nu \wedge \gamma^\alpha \wedge \gamma^\beta = -(\gamma_5\gamma^\mu\gamma^\nu\gamma^\alpha\gamma^\beta)_0. \tag{9.60c}$$

So, we we may write

$$S = (1/2)S^{\alpha\beta}\gamma_\alpha\gamma_\beta = (1/2)S^{\alpha\beta}\sigma_{\alpha\beta}, \tag{9.60d}$$

where $\sigma_{\alpha\beta}$ are the generators of rotations:

$$\sigma_{\alpha\beta} = (1/2)(\gamma_\alpha\gamma_\beta - \gamma_\beta\gamma_\alpha). \tag{9.60e}$$

9.4 Generators of Rotations in Space-Time: Intrinsic Spin

Equation 9.53, Equation 9.54a and Equation 9.60a yield

$$S\psi = (\hbar/2)\psi\gamma_2\gamma_1 \tag{9.61}$$

Interpreting this as a matrix equation, we can write, by using (9.46a, b) [4, 5]:

$$S\psi u_1 = i(\hbar/2)\psi u_1, \tag{9.62}$$

or in terms of Dirac spinor Ψ (by relation (9.47)),

$$S\Psi = i(\hbar/2)\Psi. \tag{9.63}$$

This implies that $i(\hbar/2)$ is an eigenvalue of the bivector S describing the spin. So, $i(\hbar/2)$ is a representation of the spin bivector S by one of its eigenvalues.

In the two-dimensional i-plane of vectors, the unit pseudoscalar $i = \sqrt{-1}$ is a unit bivector. It is a generator of rotations in the plane as well as the 'intrinsic' rotation represented by $\sigma_1 \wedge \sigma_2$ in the spinor i-plane. The spinor plane is not detached from the real vector plane as depicted in Figure 3.1 and Figure 3.2. There is a one-to-one correspondence between them. This feature gives geometric algebra a great advantage over matrix algebra, in which abstract spin space is detached from space-time.

If we extend from two-dimensional space to four-dimensional space-time, the spin bivectors S is defined by

$$S = (\hbar/2)R\gamma_2\gamma_1\tilde{R}, \quad i(\hbar/2) \quad \text{being an eigenvalue of } S.$$

We note the following points:

1. The algebra of the i-plane is a subalgebra of the algebra of space-time.
2. There is an automorphism of the linear space of all bivectors to itself.

Thus, in conformity with the description of bivectors in two-dimensional space, one may state that $\gamma_\alpha \wedge \gamma_\beta$ are generators of rotations in space-time. The $\sqrt{-1}$ of the Dirac theory can be interpreted geometrically as the generators of rotations in the $e_2 \wedge e_1$-plane with $(\hbar/2)\ \gamma_2 \wedge \gamma_1$ representing the intrinsic spin in real Dirac algebra in conventionally used labeling.

9.4.1 General Observations

Interpreting the γ_μ in the Riemann–Cartan space-time as vectors instead of matrices, the geometric product $\gamma_\mu\gamma_\nu$ can be understood geometrically by separating it into symmetric and antisymmetric parts [2]

$$\gamma_\mu\gamma_\nu = \gamma_\mu \cdot \gamma_\nu + \gamma_\mu \wedge \gamma_\nu, \tag{9.64a}$$

where

$$\gamma_\mu \cdot \gamma_\nu = (1/2)(\gamma_\mu\gamma_\nu + \gamma_\nu\gamma_\mu) \equiv g_{\mu\nu}, \tag{9.64b}$$

$$\gamma_\mu \wedge \gamma_\nu = (1/2)(\gamma_\mu\gamma_\nu - \gamma_\nu\gamma_\mu) \equiv \sigma_{\mu\nu}. \tag{9.64c}$$

It has been pointed out [6] that in the gauge theory of gravity as well as in supergravity, if supersymmetric action contains the graviton (e^k_μ) and the gravitino (ψ^α_μ), the affine connection is nonsymmetric because of the presence of a fermion field that is a source of torsion according to the Einstein–Cartan theory.

The equation

$$\gamma_\mu\gamma_\nu = g_{\mu\nu} + \sigma_{\mu\nu} \tag{9.65}$$

suggests that supersymmetry is inherited from it because of the simultaneous existence of the commutator (fermionic field) and anticommutator (bosonic field): $g_{\mu\nu}$ is connected with bosons and $\sigma_{\mu\nu}$ with fermions.

Next, the covariant differential of a spinor field is (see, for instance, [6])

$$D\psi = dx_k \partial_k \psi + (1/2)(\omega_{ik}\gamma^{[i}\gamma^{k]}\psi) \tag{9.66}$$

yields dilation and rotation simultaneously, which, in turn, indicate the presence of both curvature and torsion.

Furthermore, in References [7–10] it has been shown that curvature R and torsion Q may play the role of conjugate variables of the geometry, giving the commutation relation

$$[Q, R] = (\hbar G/c^3)^{-3/2} \tag{9.67}$$

9.5 Fiber Bundles and Quantum Theory vis-à-vis the Geometric Algebra Approach

According to quantum theory, a fermion does not return to its initial state by a rotation of 2π, but it takes a rotation of 4π to restore its state of initial condition. First, we schematically give the well-known fiber bundle picture of the neutron interferometer experiments [11,12] in quantum theory. This, indeed, throws light on the role of gravitation in quantum mechanics [13].

The way in which a phase difference is induced between two particle rays depends, in the first case, on a magnetic field [11] and, in the second case, on the terrestrial gravitational field [13]; the results are similar in both cases in the sense that one observes the usual peaks of interference pattern. In order to see how the magnetic field can change the phase of a spinor, one has to observe the precession of the spin vector. This, in fact, allows a novel application of geometric algebra [14].

In this context we discuss multivector algebra and show how the generalized phase shifts in the abstract neutron state space of the quantum theory can be associated with the rotation angle in the real phase space by means of multivector calculus in real space-time [15].

9.6 Fiber Bundle Picture of the Neutron Interferometer Experiment

Neutron interferometer experiments [11, 12] show what happens when neutrons are rotated 360° by a magnetic field and demonstrate how a fiber bundle can arise in quantum theory. In the neutron-rotation experiment, which in fact concerns the topology of the fiber bundle, the global structure of the fiber bundle is significant. Furthermore, the experiment demonstrates a highly counterintuitive effect whose mathematical counterpart is the one-sidedness of a Möbius strip.

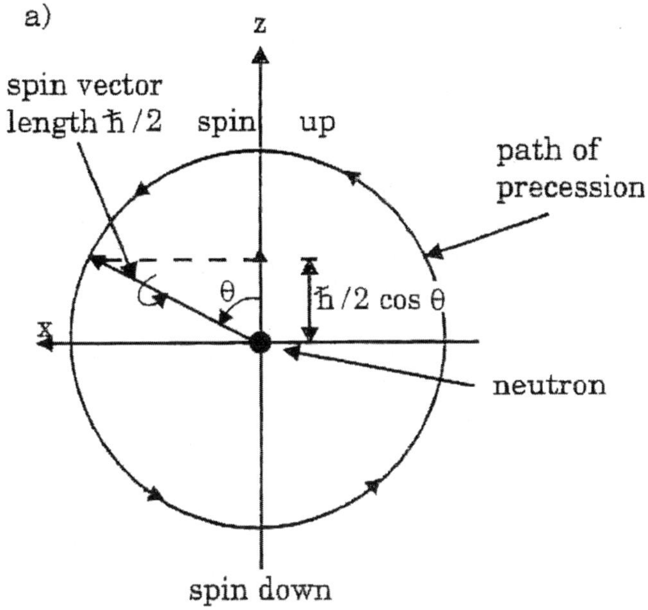

FIGURE 9.1a
Classical model of the path of precession of spin vector of a neutron. As a spin measurement along any given axis gives only the values $+\hbar/2$ or $-\hbar/2$, the classical model cannot consistently represent the geometry of the precession. (Adapted from H. J. Bernstein and A. V. Phillips, *Sci. Am.* 245, 94–109 (1981).)

The spin vector of a neutron in a magnetic field can precess, but the geometry of the precession cannot be consistently constructed by the classical model because a spin measurement assumes only the values $+\hbar/2$ (spin-up) or $-\hbar/2$ (spin-down) along any given axis (Figure 9.1a).

In quantum mechanics, precession is considered to be a manifestation as a change in the probability of finding a neutron with spin $+\hbar/2$ or with spin $-\hbar/2$. The two amplitudes that determine the probability can be considered coordinates in the abstract neutron state space with two perpendicular axes labeled "up" and "down" (Figure 9.1b). After a precession of 90° from the z-axis, the spin vector points neither up nor down (Figure 9.1a, b). If one measures the z-axis component of the vector, one finds spin-up half of the time and spin-down half of the time. The average value of the spin is zero, in agreement with the classical result. As the probabilities are equal, the probability amplitudes can be chosen to be equal; the corresponding point in neutron-state space (an abstract space) is rotated 45° from the up-axis. So, physical precession through any angle ϑ causes a generalized phase shift $\vartheta/2$, represented as a rotation in the neutron-state space (Figure 9.1b).

Now, we give the fiber bundle model (see, for instance, Reference [14]) of the phase shift as the generalized phase of the neutron spin state (Figure 9.2). The fiber bundle model of the phase shift shows the relation between the

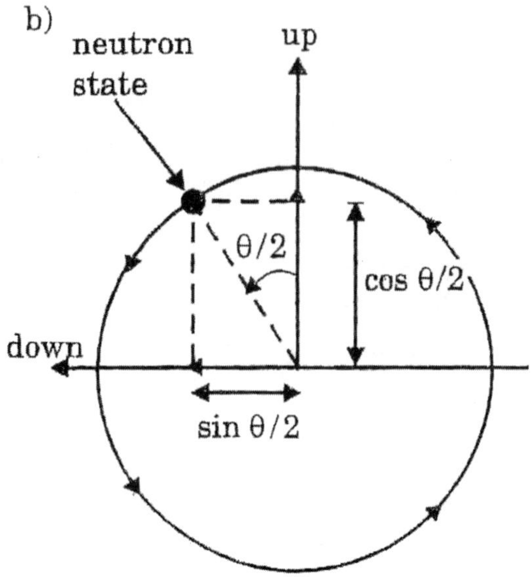

FIGURE 9.1b
Quantum mechanical picture of the neutron interferometer experiment. Two perpendicular axes labeled "up" (corresponding to the z-axis of the classical model) and "down" are constructed in the abstract neutron-state space, which is a unit circle. The coordinates in the abstract space can quite plausibly represent the two amplitudes that determine the probability of finding a neutron with spin $+\hbar/2$ and $-\hbar/2$. So, the precession of a spin vector of a neutron is manifested as a change in the probability of finding a neutron with spin-up or spin-down. Physical precession through any angle ϑ (see Figure 9.1a) can cause a generalized phase shift $\vartheta/2$, which is represented as a rotation in the neutron-state space. Thus a precession of 90° from the z-axis, where the spin vector points neither up nor down, corresponds to the point in the neutron-state space rotated through 45° from the "up" axis. (Adapted from H. J. Bernstein and A. V. Phillips, *Sci. Am.* 245, 94–109 (1981).)

angular precession of a neutron and the shift in the generalized phase of the neutron spin state. Points in the base space of the bundle represent the orientation of the spin vector of a neutron. Points in the total space represent the relative phase shifts in the neutron-state space that correspond to a given orientation. For instance, the projection map of the bundle assigns the points 45° and 225° in the total space to the point 90° in the base space. This means that generalized phase angles of 45° and 225° both correspond to an orientation of the vector 90° from the z-axis. The underlying geometric principle for the above correspondence is that the point on the phase circle moves continuously in such a way that it always remains above the point on the orientation circle. Thus, the topological structure of the bundle, together with the above geometric principle, accounts for the sign change of the neutron state as an effect of a phase reversal, the total space and the base space both being topologically equivalent to a circle. One complete rotation in the base circle must shift the generalized phase to the opposite of what it was before.

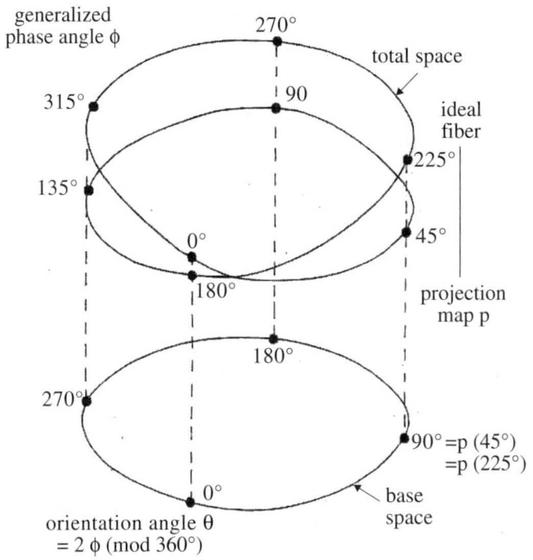

FIGURE 9.2
Fiber bundle picture of a neutron spin rotation. The correspondence between the angular precession of a neutron and the shift in the generalized phase of the neutron spin state is depicted by the fiber bundle picture of phase shifts. Points in the base space of the bundle represent the orientation of the spin vector of a neutron, whereas those in the total space represent the relative phase shifts in the neutron state space that correspond to a given orientation. As the orientation angle ϑ equals twice the generalized phase angle ϕ (modulo 360°), the projection map p of the bundle assigns the points 45° and 225° in the total space to the point 90° in the base. The total space and the base space are topologically equivalent to a circle. The projection map corresponds to the way the edge of a Möbius strip would project onto a circle at the center of the strip. (Adapted from H. J. Bernstein and A. V. Philips, *Sci. Am.* 245, 94–109 (1981).)

This is why the projection map corresponds to the way the edge of a Möbius strip would project onto a circle at the center of the strip.

9.6.1 Multivector Algebra

Now, we will use some aspects of multivector algebra (see Chapter 5) in order to show its potency in physical applications, and, in particular, through the definition of the geometric product of vectors in four-dimensional space-time, which provides a geometrical interpretation of the imaginary numbers and the reinterpretation of the Dirac equation in *real* space-time without imaginary numbers [16].

In this regard, Yu Xin [17] shows that the concept of internal and external spaces are just two different representations of a single primitive structure — the spinorial space-time; in other words, the spinor structure represents the internal abstract space and external space-time, depending on its representation.

In fact, we have seen that in order to take into account both mass and spin, it seems at first sight that we have to do with two different spaces: a real space-time where we describe the curvature, due to the mass, with tensors; and a complex space-time, where we describe torsion, due to the spin, with spinors. But this is not completely satisfactory; one would like to describe these two fundamental physical properties, mass and spin, in a unique manifold (see Reference [17]), the real space-time, and this can be done through Hestenes algebra [5,18].

In fact, for studying problems, in the early universe we have to deal both with elementary particle physics using quantum theory and with cosmology using general relativity; however, general relativity is developed in real space-time, whereas quantum theory needs a complex manifold. How can we conciliate general relativity with quantum theory? This difficulty could be overcome by describing spinors in real space-time.

For this reason we consider the geometric algebra that, with the multivector concept and the interpretation of imaginary units as generator of rotations, places tensors and spinors on the same footing: both are described in a real space-time [16].

We see that the Hestenes STA [5,18] automatically incorporates the geometric structure of space-time.

This can be done introducing, first of all, the *outer product* $a \wedge b$, which is different from the usual cross product in the sense that it has magnitude $|a||b| \sin\vartheta$ and shares its skew property $a \wedge b = -b \wedge a$, but is not a scalar or a vector; it is a *directed area*, or *bivector*, oriented in the plane containing a and b. One can visualize the outer product as the area swept out by displacing a along b with the orientation given by traversing the so-formed parallelogram first along the a vector and then along the b vector.

One can generalize this notion to products of objects with higher dimensionality or *grade* [19] in the sense that if the bivector $a \wedge b$, which has grade 2, is swept along another vector c of grade 1, one obtains the directed volume $(a \wedge b) \wedge c$, which is a trivector of grade 3. Thus, we are led to the notion of a multivector.

We are now in a position to define the geometric product: it is the sum of the inner and outer product, that is (dropping the convention of using the bold-face type for vectors),

$$ab = a \cdot b + a \wedge b. \tag{9.68}$$

Now, one can proceed to reformulate Dirac's theory in terms of space-time geometric calculus without any complex number. We can write the Dirac equation as [20, 21, 22]

$$\hbar \nabla \psi \gamma_2 \gamma_1 - (e/c) A \psi = mc \psi \gamma_0 \tag{9.69}$$

where ∇ is the four-dimensional generalization of the gradient operator, which, taking into account the metric, is

$$\nabla \equiv \gamma^\mu \partial_\mu, \quad \partial_\mu \equiv \partial / \partial x^\mu, \quad A = A_\mu \gamma^\mu = A^\mu \gamma_\mu, \tag{9.70}$$

and ψ is connected with Dirac column spinor Ψ by $\Psi = \psi u$, u being the unit column spinor.

9.6.2 Lorentz Rotations

Now, we proceed to find the Lorentz rotation as a multivector [22]. In the algebra of Dirac matrices the conditions

$$(1/2)(\gamma_\mu\gamma_\nu + \gamma_\nu\gamma_\mu) = g_{\mu\nu}I, \tag{9.71}$$

$$(1/4)Tr\gamma_\mu\gamma_\nu = g_{\mu\nu}, \tag{9.72}$$

where the $g_{\mu\nu}(\mu, \nu = 0, 1, 2, 3)$ is the space-time metric tensor and I is the unit matrix, do not determine the Dirac matrices uniquely. Any two sets of Dirac matrices $\{\gamma_\mu\}$ and $\{\gamma_\mu^{'}\}$ are related by a similarity transformation

$$\gamma_\mu^{'} = R\gamma_\mu R^{-1}, \tag{9.73}$$

where R is a nonsingular matrix. This, in fact, gives a change in representation of the Dirac matrices. In STA the geometrical requirement that γ_μ' in (9.73) must be vectors implies that they can be expressed as

$$\gamma_\mu' = a_\mu^{\ \nu}\gamma_\nu \tag{9.74}$$

This means that (9.73) must be invariant under reversion

$$\tilde{R}\gamma_\mu' = \gamma_\mu\tilde{R}, \tag{9.75}$$

where we define conjugate multivector $\tilde{M}$(reversion of M) of M of the real Dirac algebra by reversing the order of the product of all vectors of M. As it is independent of any basis of the algebra, it is an invariant type of conjugation. Now, from (9.73) and (9.75) we obtain

$$\tilde{R}R\gamma_\mu R^{-1} = \gamma_\mu\tilde{R}. \tag{9.76}$$

So, one may choose R such that

$$\tilde{R}R = 1 \quad \text{or} \quad R^{-1} = \tilde{R}. \tag{9.77}$$

Then, (9.73) assumes the form

$$\gamma_\mu' = a_\mu^{\ \nu}\gamma_\nu = R\gamma_\mu\tilde{R}, \tag{9.78}$$

which represents a Lorentz transformation from a frame $\{\gamma_\mu\}$ into a frame $\{\gamma_\mu'\}$. Furthermore, one can solve Equation 9.78 for R as a function of γ_μ and γ_μ' only. This implies that R is an even multivector (remember that an even multivector is a spinor in geometric algebra) and that every Lorentz transformation can be expressed in this form.

Equation 9.78 represents a proper Lorentz transformation (i.e., transformations continuously connected to the identity) if and only if R is an even multivector satisfying Equation 9.77 [15]. From this consideration one can write:

$$R = exp[-\phi/2], \tag{9.79}$$

where ϕ is a bivector. R is referred to as a Lorentz rotation. In fact, R is a real spinor field in STA.

We can apply the general theory of rotation in three-dimensional space as seen in the algebra of a plane [15]. The number of the parameters that define a rotation is equal to the number of the base bivectors, i.e., three. We can write any bivector of unit modulus as in, where n is the unit vector ($n^2 = 1$) that specifies the rotation axis. So, ϕ can be expressed as

$$\phi = in\,\vartheta, \tag{9.80}$$

where ϑ is the rotation angle around the n-direction. Then, the spinor Rin (9.79) can be written as

$$R = \exp(-in\vartheta/2) = \cos(\vartheta/2) - in\sin(\vartheta/2), \tag{9.81}$$

which satisfies the condition (9.77). The action of R on a vector a is written as

$$b = Ra\tilde{R} \tag{9.82}$$

This relation allows us to calculate b algebraically through the geometric product between a and n. If the rotation is infinitesimal, developing the exponential up to the first order $R = 1 - (in\vartheta/2)$, we have, from (9.82),

$$\begin{aligned} b &= [1 - (in\,\vartheta/2)]a[1 + (in\,\vartheta/2)] \\ &= a - (in\,\vartheta/2)a + a(in\,\vartheta/2) \end{aligned} \tag{9.83}$$

up to the first order. Thus, we have for the variation of a

$$\delta a = b - a = [(ain\; - \; ina)/2]\vartheta = i(a \wedge n)\vartheta \tag{9.84}$$

(remember that $(a \wedge n) = (1/2)(an - na)$, and that in Pauli algebra i commutes with all vectors indicating its similarity in character with the unit imaginary). Then δa, the variation of a, is orthogonal to both a and n. As regards the composition of rotations, we have (rotating the vector b with the spinor R′)

$$c = R'b\tilde{R}' = (R'R)a(\tilde{R}\tilde{R}') = (R'R)a(R'R)\tilde{}, \tag{9.85}$$

so that, comparing the two equations, one can see that $R'R$ is also a rotation (the transformation from a to c corresponds to the transformation $R \Longrightarrow R'R$); it represents the law of left composition of rotation group so that R′ operates on R with a left action without touching a. If with R' we perform a rotation of 360°, b stays invariant but $R \Longrightarrow -R$ (in fact, $R' = \tilde{R}' = -1$; see (9.81), and then $R'R = -R$, i.e., $R \Longrightarrow -R$). Now, either R or $-R$ transforms a to b, i.e., they represent the same rotation but obviously are two distinct elements of the algebra. The correspondence between spinors and rotations is 2 to 1. This is the well-known relation between the matrices of the unitary group SU(2) and the matrices of the orthogonal group SO(3).

Thus, we note that the well-known change of sign of a fermion spin for a rotation of 360° in quantum theory is due to the fact that R′ acts only on one side of the spinor; on the contrary, for a vector, as shown on the foregoing, it acts on both sides ($a \Longrightarrow R' a \tilde{R}'$ and then $a \Longrightarrow a'$ for $R' = -1$) and exhibits no change of sign for a rotation of 360°.

From the foregoing discussion we note that the rotation angle around the n-direction for the real spinor R in (9.81) is, in fact, the generalized phase ϕ of a neutron spin state represented by a point in the total space of the fiber bundle of a neutron spin rotation (see Figure 9.2 of Section 9.6). For further elucidation of the above consideration, we pass on to the STA in the next section (Section 9.7).

9.6.3 Conclusion

One may have a transparent picture of the earlier section if viewed from STA. As the geometric interpretation of the γ_μ of the STA is independent of the notion of Dirac spinor, the γ_μ assumes a central position in the mathematical description of all physical systems in space-time, including relativistic quantum theory.

The future-pointing timelike vector γ_o characterizes an observer's rest frame and maps the space-time bivectors $\{\sigma_k\}$ into the orthonormal basis vectors in the Pauli algebra. The γ_0 vector determines a map of any space-time vector $a = a^\mu \gamma$ as

$$a\gamma_0 = a \cdot \gamma_0 + a \wedge \gamma_o \tag{9.86}$$

Then, $a \wedge \gamma_o$ can be decomposed into the $\{\sigma_k\}$-frame and shown to represent a spatial vector relative to an observer in the γ_o-frame. This important and novel feature embodied by Equation 9.86 demonstrates that the algebraic properties of vectors in relative space are completely determined by the properties of the relativistic STA.

One may note that the unit bivector $\gamma_2\gamma_1$ plays the role of the unit imaginary $\sqrt{-1}$ in the Dirac theory. It is a generator of rotations as well as the intrinsic spin in the real spin plane represented by $\gamma_2\gamma_1$ (see References [21], [22], [23]).

At that proposal we would like to emphasize a more general expression for the Dirac matrices when viewed through the representation with vectors in geometric algebra. In fact, considering the geometric product of the $\gamma_\mu' s$, we can write [24, 25]

$$\gamma_\mu\gamma_\nu = \gamma_\mu \cdot \gamma_\nu + \gamma_\mu \wedge \gamma_\nu, \tag{9.87}$$

and then, besides the expressions (9.64b), we have

$$\gamma_\mu \wedge \gamma_\nu = (1/2)(\gamma_\mu\gamma_\nu - \gamma_\nu\gamma_\mu), \tag{9.88}$$

which represents both the generator of rotations as well as the spin orientation.

We give a very brief sketch in order to illustrate the probable importance of that argument implied by Equation 9.87 and Equation 9.88.

Consider, in fact, the Dirac equation in gravitational field in Riemann–Cartan space-time (and this is necessary if we like to describe bosons and fermions); we know that the contorsion tensor is completely antisymmetric (see, for instance, Reference [25]). In fact, if we consider the term in the Lagrangian for the Dirac equation in U_4 that contains the interaction between spinor and torsion, we find $K_{abc}\bar{\psi}\gamma^{[a}\gamma^{b}\gamma^{c]}\psi$ (see [25]) (where K_{abc} is the contorsion tensor defined through torsion tensor as $= -Q_{abc} - Q_{cab} + Q_{bca}$), and the spin density tensor is given by

$$S^{abc} = (1/\sqrt{-g})\left[\partial\sqrt{-g}\quad \mathcal{L}_m/\partial K_{bca}\right]$$
$$= -(j/4)\bar{\psi}\gamma^{[a}\gamma^{b}\gamma^{c]}\psi = S^{[abc]}, \tag{9.89}$$

that is, the spin density tensor is totally antisymmetric. The same result of the complete antisymmetry of the torsion tensor is found by Yu [17], imposing the equivalence principle on the space-time structure. Moreover, we know that the transformation law for the spinor field is written as

$$\psi'(x') = U(\Lambda)\psi(x), \tag{9.90}$$

where $U(\Lambda)$ is the usual 4×4 constant matrix representing the Lorentz transformation (the spinor indices are not written explicity). Λ is the Lorentz matrix involved in the vector Lorentz transformation in the flat tangent space $x'^i = \Lambda^i_k x^k$. The Dirac equation ($j\gamma^k\partial_k\psi - m\psi = 0$) is transformed as

$$j\gamma^i(\partial\psi'(x')/\partial x'^i) - m\psi'(x') = j\gamma^i\Lambda^k_i\partial_k U\psi - mU\psi = 0 \tag{9.91}$$

It is well known that multiplying from the left by U^{-1} and imposing on the Dirac equation form invariance under a Lorentz transformation we obtain the condition for the matrix U:

$$U^{-1}\gamma^i U = \gamma^k\Lambda^i_k \tag{9.92}$$

Considering an infinitesimal transformation

$$\Lambda_{ik} \simeq \eta_{ik} + \omega_{ik}, \tag{9.93}$$

(where $\omega_{ik} = \omega_{[ik]}$), we have

$$U = 1 + (1/2)\omega_{ik}S^{ik} \tag{9.94}$$

The ω_{ik} are six constant infinitesimal parameters, and $S^{ik}$are the generators of the infinitesimal Lorentz transformation, which, in order that (9.92) be fulfilled, must satisfy

$$S^{ik} = \gamma^{[i}\gamma^{k]} = (1/2)(\gamma^i\gamma^k - \gamma^k\gamma^i) \tag{9.95}$$

Here, we can observe that, considering the Dirac equation in Riemann–Cartan space-time, we have the relation

$$g_{\mu\nu} = (1/2)(\gamma_\mu \gamma_\nu + \gamma_\nu \gamma_\mu) = \gamma_\mu \cdot \gamma_\nu \tag{9.96}$$

where γ_μ are the Dirac vectors. Moreover, Equation 9.95 is connected with spin [17,18]. In fact, we have seen in Equation 4.5 that $\sigma_1\sigma_2 = \sigma_1 \wedge \sigma_2$ is the generator of rotation in 2-space and we know also that in 3-space, $\sigma_1\sigma_2$, $\sigma_2\sigma_3$, $\sigma_3\sigma_1$ (see Equation 9.35 and also Equation 4.7a,b,c) are connected with spin. So, we have that in space-time $\gamma_\mu \wedge \gamma_\nu$ are the generators of rotations (Equation 9.88 and Equation 9.95), and then are also connected with spin! So, writing

$$\sigma_{\mu\nu} = (1/2)(\gamma_\mu \gamma_\nu - \gamma_\nu \gamma_\mu), \tag{9.97}$$

as from the Hestenes geometric product, we have

$$\gamma_\mu \gamma_\nu = \gamma_\mu \cdot \gamma_\nu + \gamma_\mu \wedge \gamma_\nu, \tag{9.98}$$

and we can also write

$$\gamma_\mu \gamma_\nu = g_{\mu\nu} + \sigma_{\mu\nu} \tag{9.99}$$

Equation 9.99 seems to include automatically supersymmetry because we have the commutator (fermionic field) and the anticommutator (bosonic field): $g_{\mu\nu}$ is connected with bosons and $\sigma_{\mu\nu}$ is connected with fermions, and they are given simultaneously. In other words, Hestenes geometric algebra [5] with the concept of multivectors and with the precise geometrical interpretation of imaginary numbers seems to be very important; the unit imaginary j appearing in the Dirac, Pauli, and Schrödinger equations has a geometrical interpretation in terms of rotation in real space-time [5], so one has to do with real space-time, with spinors and tensors treated in a unified way (and this allows us to write Equation 9.99) (see also Reference [16]).

As a last comment, we can observe that the relation

$$i\sigma_3 = i\gamma_3\gamma_0 = \gamma_2\gamma_1 \tag{9.100}$$

shows that the phase giving the magnitude of a rotation in the spin plane (real plane in STA) and the spin describing the orientation of the spin plane are inextricably unified. This, in fact, demonstrates the well-known change of sign of a fermion spin for a rotation of 360° in quantum theory whose mathematical counterpart is the one-sidedness of the Möbius strip. So, one may state that the generalized phase ϕ of the neutron-spin state represented by a point in the total space (corresponding to the edge of a Möbius strip) of the fiber bundle of a neutron spin rotation can be described by the rotation in the Pauli algebra via the real spinor R (see Equation 9.81).

9.7 Charge Conjugation

The Dirac equation for an electron can be written in Hestenes algebra as [1,2,23] (see also Equation 9.52b)

$$\Box\psi = (m\psi\gamma_0 + eA\psi)\gamma_2\gamma_1 \qquad (\hbar = c = 1), \tag{9.101}$$

where $\psi(x)$ is the wave function, m is the mass, e is the charge of electron, and A is the electromagnetic vector potential. We discuss here certain symmetries of the spinor field $\psi(x)$ that map the field $\psi(x)$ onto itself, preserving the wave equation (9.101) or changing it in a definite and physically meaningful way.

The wave function ψ uniquely determines a frame of tangent vectors

$$J_\mu(x) = \psi\gamma_\mu\tilde{\psi} = \rho e_\mu \tag{9.102}$$

(because $\rho_0 > 0$, J_0 is a timelike vector in the forward light cone and is equivalent to the probability current density of the Dirac theory, so ρ may be interpreted as the proper probability density) at each point of space-time, and inversely, except for a factor $e^{\gamma_\mu\beta}$, the tangent vectors J_μ determine ψ. In view of the above, one can give a geometric interpretation to the transformation of ψ because any transformation of the tangent vectors can be interpreted geometrically (β is a scalar).

Symmetries of a spinor field can be interpreted geometrically as some combination of two distinct types of geometrical transformations:

1. A transformation of the tangent vectors $J_\mu(x)$ at a point x of the space-time into a new set of tangent vectors $J'_\mu(x)$ at the same point x
2. A point transformation

$$x = x^\mu\gamma_\mu \Longrightarrow x' = x'^\mu\gamma_\mu \tag{9.103}$$

wherein the tangent vectors J_μ at a point x of space-time are mapped into equivalent vectors at a different point x'.

The transformation C, known as the Charge conjugation, changes the sign of electromagnetic coupling, leaving the rest of the Dirac equation invariant. This can be achieved if we take the charge conjugation as (see [23])

$$C : \psi \Longrightarrow \psi^C = \psi\gamma_2\gamma_0 \tag{9.104}$$

or

$$C : \psi \Longrightarrow \psi^C = \psi\gamma_1\gamma_0 \tag{9.105}$$

It is easy to see that under charge conjugation (9.104) or (9.105) the Dirac equation (9.101) becomes

$$\Box\psi = (m\psi\gamma_0 - eA\psi)\gamma_2\gamma_1. \tag{9.106}$$

The above conjugation (9.104) induces a rotation of π of J_1 around the J_2 axis:

$$J_0 \Longrightarrow J_0, \quad J_1 \Longrightarrow -J_1, \quad J_2 \Longrightarrow J_2, \quad J_3 \Longrightarrow -J_3, \qquad (9.107)$$

whereas the charge conjugation (9.105) induces a rotation of π of J_2 around the J_1 axis:

$$J_0 \Longrightarrow J_0, \quad J_1 \Longrightarrow J_1, \quad J_2 \Longrightarrow -J_2, \quad J_3 \Longrightarrow -J_3 \qquad (9.108)$$

Physically, this means that in both the cases the spin remains invariant.

Moreover, under charge conjugation (9.104) or (9.105), the bilinear function of ψ changes sign:

$$\psi^C \tilde{\psi}^C = -\psi\tilde{\psi} = -\rho e^{\gamma_\mu \beta} \qquad (9.109)$$

So, the scalar part of the bilinear function of the wave function also changes sign:

$$(\psi^C \tilde{\psi}^C)_0 = -(\psi\tilde{\psi})_0 = -\rho \cos\beta \qquad (9.110)$$

and can plausibly be interpreted as the proper particle density of the spinor field ψ. So, a negative value given by (9.110) would mean the likelihood of observing an antiparticle [23].

Appendix A

The vectors $\gamma_0, \gamma_1, \gamma_2$, appearing in the Dirac equation (9.52a,b), belong to a set of arbitrarily chosen orthonormal vectors $\gamma_\mu (\mu = 0, 1, 2, 3)$. The choice of a coordinate frame $\{\gamma_\mu\}$ with γ_0, the reference frame's 4-velocity, corresponds to the standard matrix representation of the Dirac theory for which γ_0 is Hermitian and the γ_k $(k = 1, 2, 3)$ are anti-Hermitian. Furthermore, the standard matrix representation

$$\hat{\gamma}_0 = \begin{bmatrix} I & 0 \\ 0 & -I \end{bmatrix}, \quad \hat{\gamma}_k = \begin{bmatrix} 0 & -\hat{\sigma}_k \\ \hat{\sigma}_k & 0 \end{bmatrix},$$

with

$$\hat{\sigma}_1\hat{\sigma}_2\hat{\sigma}_3 = \sqrt{-1} \quad I$$

(Where I is the 2×2 unit matrix and the $\hat{\sigma}_k$ are the usual 2×2 Pauli matrices) associates the unit imaginary $\sqrt{-1}$ or the matrix representation with the bivector $\gamma_2\gamma_1$ (9.52). So, the Hermitian conjugation and the complex numbers of the standard matrix representation can be related to some intrinsic features of the Dirac equation.

References

1. D. Hestenes, *Space-Time Algebra* (Gordon and Breach, New York, 1966).
2. D. Hestenes, *J. Math. Phys.* 8, 798 (1967).
3. D. Hestenes, *Amer. J. Phys.* 39, 1013 (1971).
4. D. Hestenes, *J. Math. Phys.* 16, 556 (1975).
5. D. Hestenes and G. Sobczyk, *Clifford Algebra to Geometric calculus* (D. Reidel, Dordrecht 1984).
6. V. de Sabbata and N. Gasperini, *Introduction to Gravitation* (World Scientific, Singapore 1985), pp. 249–270.
7. V. de Sabbata, *Comm. Theor. Phys.* (India) 2, 217 (1993).
8. V. de Sabbata, *Nuovo Cimento* 107*A*, 363 (1994).
9. V. de Sabbata, "Twistor, Torsion, and Tensor-Spinors Space-time," in *Quantum Gravity*, ed. P. G. Bergmann, V. de Sabbata and H. -J. Treder (World Scientific, Singapore) pp. 80–93, 1995.
10. V. de Sabbata and L. Ronchetti, *Found. Phys.* 29,1099 (1999).
11. S. A. Werner, R. Colella, A. W. Overhauser, and C. F. Eagen, Observation of the phase shift of a neutron due to precession in a magnetic field, *Phys. Rev. Lett.* 35, 1053–1055 (1975).
12. M. P. Silverman, The curious problem of spinor rotation, *Eur. J. Phys.* 1, 116–123 (1980).
13. S. A. Werner, R. Colella, and A. W. Overhauser, Observation of gravitationally induced quantum interference, *Phys. Rev. Lett.* 34, 1472–1474 (1975).
14. H. J. Bernstein and A. V. Phillips, Fiber bundles and quantum theory, *Sci. Am.* 245, 94–109 (1981).
15. B. K. Datta, V. de Sabbata, and R. Datta, Fiber bundles and quantum theory vis-a-vis geometric algebra approach, *Nuovo Cimento*, 114*B*, 459 (1999).
16. B. K. Datta, V. de Sabbata, and L. Ronchetti, Quantization of gravity in real space-time, *Nuovo Cimento B*113, 711–732 (1998).
17. Xin Yu (A. Yu), The Ω-field theory of gravitation and cosmology, *Astrophys. Space Sci.* 154, 321–331 (1989); see also "General Relativity on Spinor-Tensor Space-time, in *Quantum Gravity*, ed. P. G. Bergmann, V. de Sabbata, and H.-J. Treder (World Scientific, Singapore, 1996), pp. 382–411.
18. D. Hestenes, *New Foundation for Classical Mechanics* (D. Reidel, Dordrecht, Holland, 1985), pp.39–64.
19. S. Gull, A. Lasenby, and C. Doran, Imaginary numbers are not real, *Found. Phys.* 23, 1175–1201 (1993).
20. C. Doran, A. Lasenby, and S. Gull, States and operators in the spacetime algebra, *Found. Phys.* 23, 1239–1264 (1993).
21. B. K. Datta, "Physical Theories in Space-Time Algebra," in *Quantum Gravity*, ed. G. Bergmann, Venzo de Sabbata and Hans-Jürgen Treder (World Sci. Singapore, 1996) pp. 54–79.
22. B. K. Datta, R. Datta, and V. de Sabbata, "Einstein field equations in spinor formalism: A Clifford-Algebra Approach," *Found. Phys. Letters* 11, 83–93 (1998).
23. B. K. Datta and V. de Sabbata, Hestenes geometric algebra and real spinor fields, in *Spin in Gravity: Is It Possible to Give an Experimental Basis to Torsion?* ed. P. G. Bergmann, V. de Sabbata, G. T. Gillies, and P. I. Pronin (World Scientific, Singapore, 1998), pp. 33–50.

24. V. de Sabbata, "Twistors, torsion, and tensor-spinors space-time, in *Quantum Gravity* ed. P. G. Bergmann, V. de Sabbata, and H. -J. Treder (World Scientific, Singapore, 1996), pp. 80–93.
25. V. de Sabbata, "Evidence for torsion in gravity?" in *Spin in Gravity: Is It Possible to Give an Experimental Basis to Torsion*?, edited by P. G. Bergmann, V. de Sabbata, G. T. Gillies, and P. I. Pronin (World Scientific, Singapore, 1998), pp. 51–85.

10

Quantum Gravity in Real Space-Time (Commutators and Anticommutators)

10.1 Quantum Gravity and Geometric Algebra

In this chapter we will develop the problem of quantum gravity and will show, in this regard, how it is important in the consideration of geometric algebra.

For this purpose we will discuss, first, the introduction of spin in the Einstein equations of general relativity. For that reason we begin by considering the Einstein–Cartan theory, i.e., general relativity plus torsion, which, from the physical point of view, means the introduction of spin in the theory of gravity.

Now, we know that the Einstein theory of gravity, that is, General Relativity, takes into account only the mass. This is good for the macroscopic body; the mass is the source of gravity in the sense that the mass is responsible for the curvature of space-time. We know also that the general theory of relativity is the best and the simplest gravitation theory that is in agreement with all experimental facts in the domain of macrophysics, including the more recent experiment on time-delay, with radar on Mercury and Venus and other sophisticated experiments in the solar system.

However, when we consider the early universe, we know that the cosmological problem is strictly connected with elementary particle physics. Then, we must pay attention to this question: when we consider together general relativity and elementary particle physics, the latter described by the quantum field theory, we are obliged to take into account not only the mass of elementary particles but also the spin. In fact, elementary particles are characterized not only by mass but also by the spin that occurs in units of $\hbar/2$. Mass and spin are two elementary and independent original concepts: as a mass distribution in a space-time is described by the energy–momentum tensor, so a spin distribution is described in a field theory by a spin density tensor. As the mass is connected with the curvature of space-time, the spin will be connected with another geometrical property of space-time so that we must consequently modify the general theory of relativity in order to connect this new geometrical property with the spin density tensor. In this way we are led to the notion of torsion. In fact, all elementary particles can be classified by means of the irreducible unitary representation of the Poincaré group and can be labeled with the translational part of the Poincaré group, whereas spin is connected with the rotational part. In a classical field theory, mass corresponds to a canonical stress–energy–momentum tensor, and spin to a canonical spin tensor. The dynamical relation between the stress–energy–momentum tensor and curvature is expressed in general relativity by the Einstein equations; one feels here a need for an analogous dynamical relation including the spin density tensor. As this is impossible in the framework of general relativity, we are forced to introduce this new geometrical property we call torsion. We can say that as the mass is responsible for curvature, spin is responsible for torsion. We now will see, from a formal point of view, in what way we must modify the general relativity theory: the main point is to assume an affine asymmetric connection instead of the symmetric connection we have in the Einstein theory (the Christoffel symbols). Torsion is, in fact, connected with the antisymmetric part of the affine connection, as we shall see.

In this way we are led to a generalization of a Riemann space-time. This generalization was proposed in 1922 [1] by Cartan. He relates the torsion tensor to the density of intrinsic angular momentum well before the introduction of the modern concept of spin. According to Trautman, "the Einstein–Cartan theory is the simplest and the most natural modification of the original Einstein theory of gravitation" [2].

From the geometrical point of view, torsion is simply the antisymmetric part of an asymmetric affine connection $\Gamma^{\mu}_{\alpha\beta}$, that is,

$$Q_{\alpha\beta\cdot}{}^{\mu} = (1/2)\left(\Gamma^{\mu}_{\alpha\beta} - \Gamma^{\mu}_{\beta\alpha}\right) \equiv \Gamma^{\mu}_{[\alpha\beta]} \tag{10.1}$$

and has a tensor character. In the presence of torsion, space-time is called the Riemann–Cartan manifold and is denoted by U_4 (the Riemann space is denoted by V_4).

We will not go into the Einstein–Cartan theory as all the development will follow close to the structure of general relativity, but we will emphasize the fundamental fact that one of the most important geometrical properties of torsion is that a closed contour in an U_4 manifold becomes, in general, a nonclosed contour in the flat space-time V_4. This nonclosure property, that is, the fact that the integral [3, 4, 5]

$$l^{\alpha} = \oint Q^{\alpha}_{\beta\gamma} dS^{\beta\gamma} \neq 0 \tag{10.2}$$

(where $dS^{\beta\gamma} = dx^{\beta} \wedge dx^{\gamma}$ is the area element enclosed by the loop) over a closed infinitesimal contour is different from zero, can be treated as defects in space-time in analogy to the geometrical description of dislocations (defects) in crystals; this can constitute a way to move toward the quantization of gravity, which means quantization of space-time itself.

We will see that with torsion we can try the quantization of gravity. In fact, to quantize gravity, we need to find some geometric objects that behave as independent variables from which we can form the uncertainty relations. For instance, in general relativity, if we consider the metric tensor $g_{\mu\nu}$ and the affine connection $\Gamma^{\rho}_{\mu\nu}$, we cannot form any relation analogous to the uncertainty relations because $g_{\mu\nu}$ and $\Gamma^{\rho}_{\mu\nu}$ are not independent variables and $\Gamma^{\rho}_{\mu\nu}$ is not a tensor; in a moment we will have a better picture of these concepts and will also see how both of these difficulties disappear if we consider torsion. That is, this problem may not arise if torsion is considered because, in this case, the antisymmetric part of the connection $\Gamma^{\alpha}_{[\beta\gamma]}$, i.e., the torsion tensor $Q^{\alpha}_{\beta\gamma}$, is a true tensorial quantity, and further, torsion Q and curvature R are two independent geometrical variables. As we have said, with torsion one can define distances in this sense. If we consider a small closed circuit and write

$$l^{\alpha} = \oint Q^{\alpha}_{\beta\gamma} dS^{\beta\gamma},$$

where $dS^{\beta\gamma} = dx^{\beta} \wedge dx^{\gamma}$ is the area element enclosed by the loop, then l^{α} represents the so-called "closure failure," i.e., torsion has an intrinsic geometric meaning: it represents the failure of the loop to close, l^{α} having the dimension of length. ($Q^{\alpha}_{\beta\gamma}$ has the dimension of inverse length, and dA is area.) We will see in the following text that with curvature and torsion we can form uncertainty relations or commutation relations, as

$$[Q, R] = i(\hbar G/c^3)^{-3/2} \tag{10.3}$$

At this point let us remember the important works of Treder and Borzeszkowski [6,7,8] who consider the Einstein–Schrödinger affine theory in which the field coordinates are the Einstein affine tensors $U^i = \Gamma^i - \Gamma^r \delta^i_r$, and conjugate momenta are given by the tensor $R^{kl}\delta^o_i$, where $\Gamma^i_{[kl]} \neq 0$ and $R_{kl} \neq R_{lk}$. So, in the purely affine theory of Einstein and Schrödinger we have (see [7,8])

$$\left[U^a, R^{kl}\delta^0_i\right] = -i\delta^{akl}_{bci}\delta^3(x),$$

where $R_{kl} \neq R_{lk}$ is again the Einstein tensor, and $U^i_{kl} = \Gamma^i_{kl} - \Gamma^r_{kr}\delta^i_l$ the Einstein's affine tensor with $U^i_{[kl]} \neq 0$.

In particular, the conclusion drawn in Reference [6] was that the inequality relations $g_{ik}L_0^2 \geq \hbar G/c^3$ and $\Gamma^i L_0^3 \geq \hbar G/c^3$ (L_0 denotes the dimension of the spatial region over which the value of g_{ik} and Γ^i_{kl} is measured) should not be expected to follow commutation rules because the quantities "field strength" and "length" appearing in these relations cannot be defined independently of each other. Therefore, these inequalities do not have the status of uncertainty relations. Moreover, Γ^i_{kl} is not a tensor. As we have said, these difficulties disappear if we use torsion and curvature because these are independent quantities and, moreover, tensors.

10.2 Quantum Gravity and Torsion

For all the reasons said in the foregoing text, from now on we will work with torsion. We would like to say that torsion may constitute a way toward quantization of gravity. In fact, we will show that introducing torsion in General Relativity, that is, considering the effect of spin and linking torsion to defects in space-time topology, we can have a minimal unit of length and a minimal unit of time.

In the relation (10.3), we know that torsion can be related to the intrinsic spin $\hbar$, and, as the spin is quantized, we can say that the defect in space-time topology should occur in multiples of the Planck length $(\hbar G/c^3)^{1/2}$, i.e.,

$$\oint Q^{\alpha}_{\beta\gamma} dx^{\beta} \wedge dx^{\gamma} = n(\hbar G/c^3)^{1/2} n^{\alpha} \tag{10.4}$$

(n is an integer, and n^{α} = unit point vector).

This is analogous to the well-known $\oint \mathrm{pdq} = n\hbar$, i.e., the Bohr–Sommerfeld relation. So, distance has been defined independently of g_{ik}. In fact, Equation 10.4 would define a minimal fundamental length, i.e., the Planck length entering through the minimal unit of spin or action $\hbar$. So, $\hbar$ has to deal with the intrinsic defect built into the torsion structure of space-time through $l^{\alpha} = \oint Q^{\alpha}_{\beta\gamma} dx^{\beta} \wedge dx^{\gamma}$.

Therefore, the Einstein–Cartan theory of gravitation should, in contrast to Einstein's General Relativity Theory, provide genuine quantum–gravity effects. We can also observe that from relation (10.4), considering the fourth component, time, can be defined in the quantum geometric level through torsion as

$$t = (1/c) \oint Q dA = n(\hbar G/c^5)^{1/2}. \tag{10.5}$$

So, torsion is essential to have a minimum unit of time $\neq 0$!

This would give us the smallest definable unit of time as $(\hbar G/c^5)^{1/2} \cong 10^{-43} s$. In the limit of $\hbar \Rightarrow 0$ (classical geometry of general relativity) or $c \Rightarrow \infty$ (Newtonian case), we would recover the unphysical $t \Rightarrow 0$ of classical cosmology or physics. So, both $\hbar$ and c must be finite to give a geometric unit for time (i.e., $\hbar \Rightarrow 0$ and $c \Rightarrow \infty$ are equivalent). The fact that $\hbar$ is related to a quantized timelike vector discretizes time. This quantum of time or minimal unit of time also correspondingly implies a limiting frequency of $f_{\max} \approx (c^5/\hbar G)^{1/2}$. This would have consequences even for perturbative QED in estimating the self energies of electrons and other particles, i.e., the self energy integral (in momentum space) taken over the momenta of all virtual photons. To make the integral converge, Feynman, in his paper on QED [8], multiplied the photon propagator k^{-2} by the ad hoc factor $-f^2/(k^2 - f^2)$, where k is the frequency (momentum) of the virtual photon. This convergence factor, although it preserves relativistic invariance, is objectionable because of its ad hoc character without any theoretical justification. Feynman considers

f to be arbitrarily large without definite theoretical basis. Here, the presence of space-time defects associated with torsion due to intrinsic spin would give a natural basis for the maximal value for $f_{\max}^2$ as (from 10.5)

$$f_{\max}^2 \approx c^5/G\hbar \approx 10^{86}, \tag{10.6}$$

(and extremely large as required by Feynman), giving the finite result (instead of ∞) for the self energy. This makes $f_{\max}$ another fundamental constant for particle physics, serving as a high frequency cut off that is not arbitrary.

We can start from these considerations (see, for instance, Reference [3]): if we wish to connect the initial and final positions of one and the same particle, we cannot avoid uncertainty associated with torsion, i.e., for a sufficiently small area element dS, uncertainty in distance between the initial and final position would be $\Delta l^\mu = Q^\mu dS$, and this would induce fluctuations in distance in the metric through $\Delta l = \sqrt{g_{\mu\nu} dx^\mu dx^\nu}$. So, what is important is not the point, themselves but the "fluctuations" in their position, i.e., the interval between them caused by a deformation of space itself through torsion. Note that plastic deformations are induced by torsion and are different from elastic deformations considered by Sacharov (which depend only on curvature). With quantized values, these fluctuations would also manifest as metric fluctuations. Because curvature causes relative acceleration between neighboring test particles, we have the momentum uncertainty related to curvature as

$$ma^\mu ds = \Delta p^\mu = mR^\mu_{\alpha\beta\gamma} \frac{dx^\alpha}{ds} dx^\beta \eta^\gamma = mcR^\mu dS, \tag{10.7}$$

(where η^γ is the separation vector between neighboring geodesics). So, as position fluctuations are given by torsion, momentum fluctuations are due to curvature, and we can interpret quantum effects (and then uncertainty principle) as consequences of space-time deformation, i.e.,

$$\Delta p^\mu \cdot \Delta x_\mu \geq \hbar$$

where

$$\Delta x^\mu = Q^\mu dS \tag{10.8}$$

and

$$\Delta p^\mu = mcR^\mu dS \tag{10.9}$$

We see that Q (torsion) and curvature (R) play the role of conjugate variables of the geometry (gravitational field), thus enabling us to write commutation relations between curvature and torsion (analogous to $[x, p] = i\hbar$) as

$$[Q, R] = i(\hbar G/c^3)^{-3/2} = iL_{Pl}^{-3} \tag{10.10}$$

Relation (10.10) can also be written as

$$\Delta Q \Delta R \geq L_{Pl}^{-3}, \tag{10.11}$$

where L_{Pl} is the Planck length.

We stress the fact that in the papers [4, 5] we are working with "non-canonical" commutation rules between ordinary space-coordinates x^i and, respectively, the torsion Q^i_{kl} and the curvature R^i_{kml} where these coordinates are not conjugate to Q and R and then these commutation rules are not founded canonically but only by geometrical hypothesis, in analogy to the (semiclassical) quantization of Bohr and Sommerfeld.

On the contrary, in Section 10.4 (see also the papers of Borzeszkowski and Treder [6, 7]) we will show that it is possible to consider torsion and curvature as canonically conjugate variables.

10.3 Quantum Gravity in Real Space-Time

Yu Xin [9] shows that the concept of internal and external spaces are just two different representations of a single primitive structure: the spinorial space-time; in other words, the spinor structure represents the internal abstract space and external space-time, depending on its representation.

In fact, we have seen that in order to take into account both mass and spin, it seems at first sight that we have to do with two different spaces: a real space-time where we describe the curvature, due to the mass, with tensors, and a complex space-time where we describe torsion due to the spin, with spinors. However, this is not completely satisfactory; one would like to describe these two fundamental physical properties, mass and spin, in a unique manifold, the real space-time, and this can be done through Hestenes algebra [10].

In other words, we can describe, at the same time, bosons and fermions: curvature and torsion must be given in the same real space-time. We know that when we introduce the Dirac equation in Riemann–Cartan space-time (and this is necessary if we would like to describe bosons and fermions), we find that the contorsion tensor is completely antisymmetric. In fact, if we consider the term in the Lagrangian for the Dirac equation in U_4 that contains the interaction between spinor and torsion, we find $K_{abc}\bar{\psi}\gamma^{[a}\gamma^b\gamma^{c]}\psi$ (see [11]) (where K_{abc} is the contorsion tensor defined through torsion tensor as $= -Q_{abc} - Q_{cab} + Q_{bca}$), and the spin density tensor is given by

$$S^{abc} = (1/\sqrt{-g})[\partial\sqrt{-g}\mathcal{L}_m/\partial K_{bca}] = -(i/4)\bar{\psi}\gamma^{[a}\gamma^b\gamma^{c]}\psi = S^{[abc]}, \tag{10.12}$$

that is, the spin density tensor is totally antisymmetric. At this point I would like to anticipate that Xin Yu [9] had found that this result could be achieved directly from the equivalence principle (see preprint in Reference [9]).

Moreover, we know that the transformation law for the spinor field is written as

$$\psi'(x') = U(\Lambda)\psi(x), \tag{10.13}$$

where $U(\Lambda)$ is the usual 4×4 constant matrix representing the Lorentz transformation (the spinor indices are not written explicity). Λ is the Lorentz matrix involved in the vector Lorentz transformation in the flat tangent space $x'^i = \Lambda^i_k x^k$.

Dirac equation ($i\gamma^k \partial_k \psi - m\psi = 0$) is transformed as

$$i\gamma^i(\partial\psi'(x')/\partial x'^{\,i}) - m\psi'(x') = i\gamma^i \Lambda^k_i \partial_k U\psi - mU\psi = 0 \tag{10.14}$$

It is well known that multiplying from the left by U^{-1} and imposing on the Dirac equation to be invariant in form under a Lorentz transformation, we obtain the condition for the matrix U

$$U^{-1}\gamma^i U = \gamma^k \Lambda^i_k \tag{10.15}$$

Considering an infinitesimal transformation

$$\Lambda_{ik} \simeq \eta_{ik} + \omega_{ik} \tag{10.16}$$

(where $\omega_{ik} = \omega_{[ik]}$), we have

$$U = 1 + (1/2)\omega_{ik} S^{ik}. \tag{10.17}$$

The ω_{ik} are six constant infinitesimal parameters and S^{ik} are the generators of the infinitesimal Lorentz transformation, which, in order that (10.15) be fulfilled, must satisfy

$$S^{ik} = \gamma^{[i}\gamma^{k]} = (1/2)(\gamma^i\gamma^k - \gamma^k\gamma^i) \tag{10.18}$$

Here, we can observe that considering the Dirac equation in Riemann–Cartan space-time we have the relation

$$g_{\mu\nu} = (1/2)(\gamma_\mu\gamma_\nu + \gamma_\nu\gamma_\mu), \tag{10.19}$$

where γ are the Dirac matrices. However, we have also

$$\sigma_{\mu\nu} = (1/2)(\gamma_\mu\gamma_\nu - \gamma_\nu\gamma_\mu) \tag{10.20}$$

We see that Equation 10.18 and Equation 10.20 are formally identical, and it seems that we can pass from one to another with the help of Hestenes algebra [10].

We also have

$$\gamma^\mu\gamma^\nu = g^{\mu\nu} + \sigma^{\mu\nu} \tag{10.21}$$

Equation 10.21 seems to include automatically supersymmetry because we have the commutator (fermionic field) and the anticommutator (bosonic field):

$g_{\mu\nu}$ is connected with bosons and $\sigma_{\mu\nu}$ is connected with fermions, and they are given simultaneously. Notice that Equation 10.18 follows directly from the Hestenes geometric product

$$\gamma_\mu \gamma_\nu = \gamma_\mu \cdot \gamma_\nu + \gamma_\mu \wedge \gamma_\nu \tag{10.22}$$

where

$$\gamma_\mu \cdot \gamma_\nu = g_{\mu\nu} \tag{10.23}$$

and

$$\gamma_\mu \wedge \gamma_\nu = \sigma_{\mu\nu} \tag{10.24}$$

Moreover, the infinitesimal variation of a spinor under the Lorentz transformation is

$$\delta\psi = \psi' - \psi = (1/2)\omega_{ik} S^{ik} \psi = (1/2)\omega_{ik} \gamma^{[i} \gamma^{k]} \psi \tag{10.25}$$

The covariant differential for a spinor field is

$$D\psi = dx^k \nabla_k \psi = \psi'(x_2) - \psi(x_1), \tag{10.26}$$

which can be written as

$$D\psi = \psi(x_2) - \psi(x_1) - [\psi(x_2) - \psi'(x_2)], \tag{10.27}$$

where we have separated the term due to translation, $\psi(x_2) - \psi(x_1)$ from the part relative to a local rotation of the tetrad $\psi'(x_2) - \psi(x_2)$. It seems that from Equation 10.27 we have, simultaneously, curvature and torsion (or, in terms of Hestenes algebra, dilation and rotation [10]).

Considering the geometric algebra in the four-dimensional space-time (see Chapter 9) and noticing that in Dirac theory we have a totally antisymmetric spin density, we are in a position to introduce the torsion trivector

$$Q = Q^{\alpha\beta\gamma} \gamma_\alpha \wedge \gamma_\beta \wedge \gamma_\gamma \tag{10.28}$$

as element of STA, where $\{\gamma_\alpha\}$ are the base vectors for which we have

$$\gamma_\alpha \gamma_\beta = g_{\alpha\beta} + \gamma_\alpha \wedge \gamma_\beta. \tag{10.29}$$

Moreover, given the curvature bivector

$$\Omega^{\alpha\beta} = (1/2) R^{\alpha\beta\mu\nu} \gamma_\mu \wedge \gamma_\nu, \tag{10.30}$$

one can form the curvature trivectors

$$R^\alpha = \Omega^{\alpha\beta} \wedge \gamma_\beta. \tag{10.31}$$

We have seen that torsion and curvature are to be considered as conjugate variables, and now we are in position to write Dirac's equation in a real form (see, for instance, Reference [12] and Reference [13]).

We have the trivectors Q and R^α. Consider now the antisymmetric part of the geometric product

$$[Q, R^\alpha] = (1/2)(QR^\alpha - R^\alpha Q) \tag{10.32}$$

This type of product between two trivectors gives a bivector for example, it is easy to verify

$$[\gamma_0 \wedge \gamma_1 \wedge \gamma_2, \quad \gamma_0 \wedge \gamma_1 \wedge \gamma_3] = \gamma_2 \wedge \gamma_3, \tag{10.33}$$

remembering that $\gamma_0\gamma_1\gamma_2 = i\gamma_3$ where i indicates the pseudoscalar unit. In the language of geometric algebra, the imaginary unit of complex numbers is substituted by a bivector; then we can have commutation relations of canonic type from Equation 10.28, Equation 10.30, Equation 10.31, and Equation 10.32:

$$\begin{aligned}[Q, R^a] &\equiv (1/2)(QR^\alpha - R^\alpha Q)\\ &= (1/2) = [Q^{\mu\nu\beta}\gamma_\mu \wedge \gamma_\nu \wedge \gamma_\beta, \quad R^{\alpha\tau\nu\beta}\gamma_\nu \wedge \gamma_\beta \wedge \gamma_\tau]\\ &= (1/4)(Q^{\mu\nu\beta}R^{\alpha\tau}_{\nu\beta} - R^{\alpha\mu}_{\nu\beta}Q^{\tau\nu\beta})\gamma_\mu \wedge \gamma_\tau = \gamma_2 \wedge \gamma_1 L_{Pl}^{-3}\end{aligned} \tag{10.34}$$

for every α, where, the first member being a bivector, the second member also is a bivector, coherently with the fact that the imaginary unit in Dirac equation is substituted by the bivector $\gamma_2 \wedge \gamma_1$ (see, for instance, Reference [4]).

As we must be in a spin plane, we have to consider the case of $\mu, \tau = 1, 2$.

In order that the commutation relation may have this form, the following six conditions should be satisfied, considering that the left-handed member of the commutation relation (10.34) is the summation of six bivector parts, (such as $\gamma_1 \wedge \gamma_2, \quad \gamma_2 \wedge \gamma_3, \quad \gamma_3 \wedge \gamma_1, \quad \gamma_0 \wedge \gamma_1, \quad \gamma_0 \wedge \gamma_2, \quad \gamma_0 \wedge \gamma_3$):

$$(1/2)Q^{1\nu\beta}R^{\alpha 2}_{\nu\beta}\gamma_{[1}\gamma_{2]} = \gamma_1 \wedge \gamma_2 L_{Pl}^{-3}. \tag{10.34a}$$

$$Q^{\mu\nu\beta}R^{\alpha\tau}\gamma_{[\mu}\gamma_{\tau]} = 0. \tag{10.34b}$$

Equation 10.34a corresponding to the bivector $\gamma_1 \wedge \gamma_2$ gives one relation, and Equation 10.34b corresponding to the bivectors

$$\gamma_2 \wedge \gamma_3, \quad \gamma_3 \wedge \gamma_1, \quad \gamma_0 \wedge \gamma_1, \quad \gamma_0 \wedge \gamma_2, \quad \gamma_0 \wedge \gamma_3$$

gives five relations.

In conclusion, the six conditions represent a "choice of gauge" with respect to the local Lorentz rotations, according to the fact that the choice of the spin plane is arbitrary.

10.4 A Quadratic Hamiltonian

In order to give a stronger basis to the consideration of torsion and curvature as canonical conjugate variables, we try to see if it is possible to improve the theory introducing a quadratic Hamiltonian function of torsion and curvature. We use as conjugate variables the torsion and curvature trivectors because the first Bianchi identity (see [19]) guarantees that it is different from zero only in the presence of torsion. Because the commutation relations between Q and R^a are defined in geometric algebra, we need to modify the usual Lagrangian field theory.

We start from the results of Reference [15]: the field equation for the multivector ψ can be written in the manifestly invariant way:

$$\nabla\left(\frac{\partial\mathcal{L}}{\partial(\nabla\psi)}\right) = \frac{\partial\mathcal{L}}{\partial\psi}, \tag{10.35}$$

where the gradient operator $\nabla = \gamma_\mu \partial^\mu$ acts as the geometric product.

The presence of ∇ or, in other words, the peculiarity of the intrinsic geometric calculus, suggests that we can define the conjugate momentum field as

$$\Pi = \frac{\partial\mathcal{L}}{\partial(\nabla\psi)} \tag{10.36}$$

and the Hamiltonian

$$H = (\nabla\psi)\Pi - \mathcal{L} \tag{10.37}$$

This is different from the usual $\mathrm{H} = \Pi\dot{\psi} - \mathcal{L}$ formula, but, following Equation 10.35 for the Lagrangian, we propose also a modified Hamiltonian that is manifestly invariant, such as Equation 10.35, so we cannot work with $\partial\psi/\partial x^0$ because it depends on the choice of the time-coordinate x^0. So, $\Pi = \frac{\partial\mathcal{L}}{\partial(\nabla\psi)}$ is the natural choice; in fact, Equation 10.35 generalizes the Lagrangian equation $(d/dt)(\partial\mathcal{L}/\partial\dot{q} - \partial\mathcal{L}/\partial q) = 0$, simply substituting d/dt with ∇.

Notice that Equation 10.35 allows for vectors, tensors, and spinors variables to be handled in a single equation, a considerable unification (see [15]).

So, we have the Hamilton equations

$$\begin{cases} \nabla\psi = \partial H/\partial\Pi \\ \nabla\Pi = -\partial H/\partial\psi \end{cases}. \tag{10.38}$$

For example, we have the Maxwell equations in vacuum, taking $\mathcal{L} = F^2/8\pi$, where the vector potential $A = A^\mu\gamma_\mu$ and the bivector $F = \nabla\wedge A$ are conjugate variables (remembering that the geometric product gives $F = \nabla A = \nabla\cdot A + \nabla\wedge A$ and Lorentz condition $\nabla\cdot A = 0$).

Moreover, because we use the curved U_4 manifold in agreement with the minimal coupling principle, we substitute the covariant operator D for ∇

defining

$$D \equiv \nabla + \gamma_a[\omega^a,], \tag{10.39}$$

where the $\{\gamma_a\}$ is an orthonormal frame of tetrads, (for which

$$\gamma_a \gamma_b = \eta_{ab} + \gamma_a \wedge \gamma_b), \tag{10.40}$$

(related with the coordinate system $\gamma_\mu = e^a_\mu \gamma_a$)
and $\omega^a = (1/2)\omega^{abc}\gamma_b \wedge \gamma_c$ is the connection bivectors.
If in Equation 10.35 we put D instead of ∇, we have

$$D\left(\frac{\partial \mathcal{L}}{\partial(D\psi)}\right) = \frac{\partial \mathcal{L}}{\partial \psi}.$$

We remember that in the tetrad basis the connection is

$$\omega^{abc} = \omega^{a[bc]} = C^{cba} - C^{bac} - C^{acb} - K^{abc} \tag{10.41}$$

(where $C_{ab}^{\ \ c} = e^\mu_a e^\nu_b \partial_{[\mu} e_{\nu]}^{\ c}$, the Ricci connection coefficients, give the Riemannian part, and K^{abc} is the contorsion tensor that is $= -Q^{abc}$ because, in this case, we are considering the totally antisymmetric torsion).

Further, it is possible to define the covariant derivative in the direction of a vector v putting

$$D_v \equiv v \cdot D = v^a \partial_a + v^a [\omega_a, \quad] \tag{10.42}$$

In geometric algebra the commutation product with bivector leaves unchanged the grade of multivectors [8], and then, in particular, we have $(\gamma_a \cdot D)\gamma_b = \omega_{ab}\gamma_c$ as we expect. We can separate the interior and exterior covariant derivative, putting $D \equiv D_I + D_E$, where

$$D_I = \nabla \cdot + \gamma_a \cdot [\omega^a, \quad] \tag{10.43}$$
$$D_E = \nabla \wedge + \gamma_a \wedge [\omega^a, \quad], \tag{10.44}$$

which, respectively, through the inner and outer product, lowers and raises the grade of the multivector on which they operate.

The D_E operator acts like the exterior covariant derivative in the language of exterior forms, so we have the Cartan structure equations

$$D_E \gamma_a = Q^{bc}\gamma_b \wedge \gamma_c \equiv Q_a \tag{10.45}$$
$$D_E\left(\omega^{\mathrm{iba}}\gamma_i\right) = (1/2)R^{\mathrm{cdab}}\gamma_c \wedge \gamma_d \equiv \Omega^{ab} \tag{10.46}$$

and the Bianchi identities

$$D_E Q^a = \Omega^{ab} \wedge \gamma_b \equiv R^a \tag{10.47}$$
$$D_E \Omega^{ab} = 0. \tag{10.48}$$

Now we are able to introduce the following Lagrangian in order to have the usual expression for conjugate variables in a quantum theory:

$$\mathcal{L} = (\hbar c/2) R^a R_a - (c^4/2G) Q^a Q_a \tag{10.49}$$

$$\mathcal{L} = (\hbar c/2)\left(R^a R_a - Q^a Q_a/L_{Pl}^2\right), \tag{10.50}$$

where Q^a and R^a are the bivectors and trivectors,

$$Q^a = Q^{a\mu\nu}\gamma_\mu \wedge \gamma_\nu; \quad R = R^{a\mu\nu\delta}\gamma_\mu \wedge \gamma_\nu \wedge \gamma_\delta$$

respectively, formed by means of the base vectors γ_μ, satisfying the relation

$$\gamma_\alpha \gamma_\beta = g_{\alpha\beta} + \sigma_{\mu\nu} \tag{10.22}$$

See also Equation 10.21 and Reference [5], where it is argued that Equation 10.22 and also the fact that the canonical conjugate variables Q^a and R^a are connected with the commutator and anticommutator rules seems to indicate that one can treat simultaneously fermions fields as well a bosonic field.

In agreement with the Bianchi identities, the momenta conjugate to the torsion bivectors are the trivectors

$$\Pi^a = \partial\mathcal{L}/\partial(DQ^a) = \hbar c R^a \tag{10.51}$$

(in natural unit $\hbar = c = 1$, we can write Π^a or R^a without distinction). The Hamiltonian results:

$$H = DQ^a \Pi_a - \mathcal{L} = (\hbar c/2)\left(R^a R_a + Q^a Q_a/L_{Pl}^2\right) \tag{10.52}$$

Then we have the Hamilton equations

$$\begin{cases} DQ^a = \partial H/\partial R^a = R^a \\ DR^a = -\partial H/\partial Q^a = -Q^a/L_{Pl}^2 \end{cases} \tag{10.53}$$

that we can write

$$\mathrm{DD}Q^a = -Q^a/L_{PL}^2, \tag{10.54}$$

or, equivalently for R^a,

$$\mathrm{DD}R^a = -R^a/L_{Pl}^2 \tag{10.55}$$

Separating in Equations 10.53 the exterior and interior parts of different grades, we have the identities

$$\begin{cases} D_E Q^a = R^a \\ D_E R^a = 0 \end{cases} \tag{10.56}$$

(the second equation is a consequence of Equation 10.48 when the torsion is totally antisymmetric) and the field equations

$$\begin{cases} D_I Q^a = 0 \\ D_I D_E Q^a = -Q^a/L_{Pl}^2 \end{cases} \tag{10.57}$$

10.5 Spin Fluctuations

Lagrangians quadratic in curvature and torsion are nothing new (see, for instance, Reference [16–19]), but there is an interesting reason supporting the choice made here of the Hamiltonian density (10.52): it seems to describe directly the quantum fluctuations at the Planck scale of space-time points as harmonic oscillators. In fact (see Equation 10.8 and Equation 10.9) we can take $R^a dS \approx \Delta p^a/m_{Pl}c$ and $Q^a dS \approx \Delta x^a$ as uncertainties, and taking $(dS)^2 L_{Pl}^{-1} \simeq dx^3$, we have $H dx^3 \simeq (\Delta p)^2/2m_{Pl} + (1/2) m_{Pl}\omega_{Pl}^2 (\Delta x)^2$, that is, just the Hamiltonian of a harmonic oscillator with Planck frequency $\omega_{Pl} = c/L_{Pl}$ and Planck mass $m_{Pl} = \hbar/L_{Pl}c$ (notice that we always consider energy fluctuation at the Planck scale).

The wave Equation 10.54 for torsion — formally a Klein–Gordon type equation for a m_{Pl} particle — could describe the propagation of the fluctuation of the background geometry due to the torsion itself; in fact, this equation is not linear because the connection also contains the torsion and then acts on itself.

In this quantum physical context, propagating torsion is not in contradiction with the Einstein–Cartan theory — for which it is zero in vacuum — because we have to do with vacuum polarization as the source through spin fluctuations.

The quadratic Lagrangian $\mathcal{L}$ is incomplete for a gravitation theory because it does not give the Einstein equations, and we know that a quantum theory of gravity must be reduced to the classical theory when one considers gravitation far away from the Planck scale.

In other words, the problem with such quadratic Lagrangian is that one has to add to them the Einstein–Hilbert Lagrangian in order to obtain, in the Newtonian approximation, the Laplace equation.

The Lagrangian (10.50) completed by the Einstein–Hilbert term (the usual linear Lagrangian $-\mathring{R}/2\chi$) is

$$\mathcal{L}' = \mathcal{L} - \mathring{R}/2\chi,$$

i.e.,

$$\mathcal{L}' = (\hbar c/2)\left(R^a R_a - Q^a Q_a/L_{Pl}^2\right) + \mathring{R}/2\chi \tag{10.58}$$

where $\mathring{R}$ is the scalar curvature related to the Riemannian part of the connection and $\chi = 8\pi G/c^4$ (i.e., $L_{Pl}^2 = \hbar c \chi/8\pi$), reminding us of an early approach to quantum electrodynamics considered by Heisenberg, Euler, and Kockel [20].

To describe quantum corrections to classical electrodynamics, these authors had added the Larmor Lagrangian $\Lambda_0 = F_{ik}F^{ik}$ of the classical theory by quadratic invariants such as $(F_{ik}F^{ik})^2$.

It was shown that, to some extent, these corrections represent a pheno-meno-logical approximation to the exact quantum theory (see [21]). At first sight, the Lagrangian (10.58) is a gravitational analog of the electrodynamic Heisenberg–Euler–Kockel *Ansatz*. However, in contrast to the electrodynamics, the quadratic gravitational terms are not formed from only those field quantities that occur in the R-item of the Lagrangian; besides the metric, there the connection comes into the game. This is in accordance with the fact we have mentioned — that there are no genuine quantum effects in a purely metric theory. Thus, quadratic terms formed only from the metric cannot be related to quantum effects. Genuine quantum terms must contain additional fields such as torsion.

The theory given by the Lagrangian (10.58) couples two types of gravity,— metric and torsional gravity — in which the first one dominates at large and the second at small distances. Its canonical structure should turn out to be of such a type that it recovers canonical quantum GRT (which is essentially classical in the above-discussed sense) for weak fields and canonical quantum gravity given by the Lagrangian (10.50) for strong fields. In terms of a cosmological scenario, in early universe one had a strong gravity era described by Riemann–Cartan geometry that later goes over into a weak gravity era described by Riemann geometry. Strong gravity can also dominate in superdense matter, e.g., in neutron stars. To found an exact theory of quantum gravity on this basis one had to construct a Hamiltonian formalism starting from the Lagrangian (10.58).

Now, with the tetrad field (10.40) as variables, we can write

$$-\mathring{\mathrm{R}} = \mathring{\mathrm{R}}^{ab}\gamma_a\gamma_b - (1/2)\gamma_a\gamma^a\mathring{\mathrm{R}}, \tag{10.59}$$

where $\mathring{\mathrm{R}}^{ab}$ is the Ricci tensor (and $\gamma_a\gamma^a = 4$), but here it is a function of γ^a. In that way, neglecting the quadratic terms of $\mathcal{L}'$, we can have the Einstein equations using Equation 10.59:

$$(1/2)(\partial\mathring{\mathrm{R}}/\partial\gamma_a) = \mathring{\mathrm{R}}^{ab}\gamma_b - (1/2)\gamma^a\mathring{\mathrm{R}} = \chi\,\partial\mathcal{L}_m/\partial\gamma_a, \tag{10.60}$$

where $\mathcal{L}_m = \mathcal{L}_m(\psi, \gamma^a, Q^a)$ is the matter Lagrangian, and $\partial\mathcal{L}_m/\partial\gamma_a$ defines the energy-momentum tensor $T^a = T^{ab}\gamma_b$ (see Equation 10.35 with D in place of ∇, and Equation 10.45).

Moreover, we can have the Einstein–Cartan equations simply taking γ_a and Q_a as variables and neglecting in $\mathcal{L}'$ only the term quadratic in curvature:

$$G^a = \chi T^a, \tag{10.61}$$

$$Q^a = \chi\,\partial\mathcal{L}_m/\partial Q_a, \tag{10.62}$$

where $G^a = (R^{ab} + \eta^{ab} R)\gamma_b$ is the nonsymmetric Einstein tensor, and $\partial\mathcal{L}_m/\partial Q_a$ defines the spin density of the matter field as torsion source.

To give, Equation 10.61 and Equation 10. 62, note that [19]

$$R_{(ab)} = \mathring{R}_{ab} + Q_{\text{acd}} Q_b^{cd} = \mathring{R}_{ab} + Q_a \cdot Q_b, \tag{10.63}$$

with Q_{abc} totally antisymmetric, i.e., $Q_{\text{abc}} = Q_{[\text{abc}]}$ as in our theory, and by calculation,

$$R_{[ab]}\gamma^b = D_I Q_a. \tag{10.64}$$

Equation 10.64 implies $R = R^a_a = \mathring{R} + Q_a \cdot Q^a$ therefore,

$$\mathcal{L}' = -R/2\chi.$$

Consider now Equation 10.54 or Equation 10.57 without executing explicity all the geometric products (inner, outer, commutator); we can achieve qualitatively important results that reveal the physical content of the equation. We write the operator D as the sum of three parts: $D = \nabla + C + Q$, i.e., differential (flat), Riemannian, and torsionic parts, respectively [see Equation 10.39 and Equation 10.41]. We have

$$(\nabla + C + Q)(\nabla + C + Q)Q = -Q/L^2_{Pl} \tag{10.65}$$

and, evaluating all different terms, we have equations of the type

$$\begin{aligned}&\Box Q + (\nabla C)Q + C(\nabla Q) + C^2 Q + C Q^2 + (\nabla Q)Q\\&+Q\nabla Q + QCQ + Q^3 + Q/L^2_{Pl} = 0\end{aligned} \tag{10.66}$$

(where $\Box = \nabla^2$).

At the lower grade of approximation this gives for quantum propagating torsion the Klein–Gordon equation in Minkowski space-time:

$$\Box Q + Q/L^2_{Pl} = 0, \tag{10.67}$$

and we expect plane wave solutions

$$Q = Q_0 \cos[(\omega_{Pl}/c)k \cdot x], \tag{10.68}$$

where $\omega_{Pl}/c = 1/L_{Pl}$, and k is the wave vector ($k^2 = 1$).

Now, we consider the first equation that we can have that is different from the Klein–Gordon type, i.e., we neglect in Equation 10.66 the Riemannian part (i.e., C) and Q^3:

$$\Box Q + Q\nabla Q + (\nabla Q)Q + Q/L^2_{Pl} = 0. \tag{10.69}$$

If we pretend that plane wave solutions are still valid, a condition is needed; moreover, we note that the term with derivatives of Q is shifted in phase of $\pi/2$ with respect to Q, so $\nabla Q = 0$ implies

$$k \cdot x/L_{Pl} = n\pi \qquad (n \text{ integer}). \tag{10.70}$$

With $k^0 = 1$ (Q only time dependent) we have the quantization of time $t = n\tau_{Pl}$, and then (if we consider the light motion), also the quantization of distances $l = nL_{Pl}$ as we have already suggested introducing space-time defects due to torsion. In fact, taking $\Delta l \approx Q\Delta S$ in Equation 10.67, we have just the harmonic oscillator equation

$$\Delta\ddot{l} + \omega_{Pl}^2 \Delta l = 0 \tag{10.71}$$

that describes the fluctuations of space-time points and its confinement at the Planck length scale.

This is analogous to the quantum concept of Zitterbewegung as fluctuation of the order of the Compton length of the position of the electron (see Reference [4]). Like torsion, Zitterbewegung seems strictly related to the spin, so we can consider ω_{Pl} in Equation 10.71 as the analogous of the Compton frequency $\omega_c = c/r_c$ see in particular Equation 9.3 in Reference [4], which reads:

$$\Delta\theta/\Delta x^0 = RdS/Q^0 dS = 1/r_c, \tag{10.72}$$

r_c being the Compton radius, showing that the frequency ω can be interpreted as the ratio between these two incertitudes, i.e., the quantum length $r_c$can be considered as the ratio between torsion and curvature, and this fact suggests again that Q and R can be considered as conjugate variables. Notice also that as the torsion is responsible for the fluctuation of the position $\Delta x^\mu \cong Q^\mu dS$, the curvature is responsible for the defect angle $\Delta\vartheta \cong \mathrm{RdS}$ (see Equation 10.51 in Reference [20]). In other words, torsion, as we have seen, can be treated as defects in space-time in analogy to the geometrical description of "dislocations" in crystals whereas curvature can be treated as defects in angles in analogy to "disclination," so that we can write:

$$\omega_{Pl}/c = 1/L_{Pl} \tag{10.73}$$

Note that the modified harmonic oscillator equation corresponding to Equation 10.69,

$$\Delta\ddot{l} + (1 + \Delta\dot{l}/c)\omega_{Pl}^2 \Delta l = 0, \tag{10.74}$$

leads us to the same consideration and implies quantization of time.

One could consider Equation 10.71 just as the limit condition that in quantum mechanics problems gives the quantization of energy, but here we work on space-time geometry itself. Moreover, Equation 10.71 implies that, at these points, Q fluctuates between $+Q_0$ and $-Q_0$ (see Equation 10.69). This fact, in agreement with the Cartan theory, can be interpreted as spin fluctuation between $\pm\hbar/2$ and could explain why spin can only have these two values. Then, the propagation of this torsion waves gives a background of quantized space-time and spin, and is rich in consequences that are to be investigated.

In this regard we can observe that spin fluctuation is in agreement with the Pauli exclusion principle. The point is that fluctuations of the order of the Planck length render undistinguishable the extremes of fluctuation and, introducing the spin, this means that we associate the spin to the Q-wave so that we have that close extremes have opposite spin (in agreement with Pauli principle); in fact, near spins (every wavelength) are ever opposite. Then, the mean value of Q over one period is always null, but this is not true for Q squared; it is typical for all wave phenomena, and we know that the square of amplitude is related to the energy. These considerations, together with the geometric identity (10.63) suggest that the torsionic part of the curvature could be converted into the Riemannian part connected with the energy-momentum tensor through the Einstein equation

$$R_{ab} = Q_a \cdot Q_b = \chi T_{ab} \tag{10.75}$$

In fact, $Q_a \cdot Q_b$ can be interpreted as spin–spin interaction energy:

$$Q_a \cdot Q_b / \chi = T_{ab} \tag{10.76}$$

This means that where there is torsion there is also curvature, and so, mass. It is clear that Equation 10.75 represents the self-interaction energy of the field Q but, because of the quantized values of spin, we can consider, as fundamental objects, the pairs of spins (with interaction $S_1 \cdot S_2$) to describe better what happens at the Planck scale in our picture of space-time and matter. As we will see, this leads us to new topics about the uncertainty relation between energy and time.

We take a volume V small enough inside a particle (a nucleon) because we need to consider a particle as an extended body (notice that $L_{Pl} \sim 10^{20}$ times smaller than the typical hadron dimension $r \sim 10^{-13} cm$). Using Equation 10.75 and Equation 10.76 we can write, for the energy density ϵ,

$$\epsilon = \Delta E / \Delta V = Q_0 \cdot Q_0 / \chi \simeq L_{Pl}^2 \hbar c n^2, \tag{10.77}$$

where $Q_0 \simeq L_{Pl}^2 n$, and n is the number of spins per unit volume: $n = N/\Delta V$. It follows that

$$\Delta E \Delta V \simeq L_{Pl}^2 \hbar c N^2, \tag{10.78}$$

or, if we consider the volume that contains only one spin ($n = 1/\Delta V$),

$$\Delta E \Delta V \simeq L_{Pl}^2 \hbar c, \tag{10.79}$$

where ΔE is the energy inside ΔV.

Now, if we take two interacting spins separated by a distance Δd (where d is less than the Compton length but greater than the Planck length), because L_{Pl} is the minimal length and the time Δt needed to connect them causally is $\Delta d/c$, then putting $\Delta V = L_{Pl}^2 c \Delta t$ in Equation 10.79, we find $\Delta E \Delta t \simeq \hbar$ (remember that we are always in the situation where the spins are anti-parallel), i.e., the well-known uncertainty relation. We give this interpretation: if the spins interact for a time Δt, there is a fluctuation in energy ΔE that is "virtual"; if the interaction is stationary, we have inside ΔV an energy that can constitute a real particle (see Equation 10.79). For example, we consider the nucleon mass and suppose that its energy can be made by N parts ΔE, which obey Equation 10.79: putting $E = N\Delta E$ and $r^3 = N\Delta V$, we find $N \simeq 10^{20}$ and $n = N/r^3 = 10^{59} cm^{-3}$. This value gives very high torsion ($Q = L_{Pl}^2 n$), but if we consider the wave equation for Q inside the nucleon as a whole, then the just mentioned spin fluctuations occur, for which $< Q >= 0$ and $< Q^2 > /\chi \simeq \epsilon \simeq mc^2/r^3$. Note that N is just the ratio r/L_{Pl}; this seems in agreement with wave interpretation, i.e., there is one spin for every wavelength L_{Pl} inside r.

In other words, we can say that with spin fluctuations one can determine the mass of the nucleon through the number $N \simeq 10^{20}$, which represents, geometrically, the ratio between Compton wave length and Planck length. As Q is confined inside the nucleon, we can take as starting point the number $N \simeq 10^{20}$ arriving in that way to the determination of the mass of a nucleon.

10.6 Some Remarks and Conclusions

We have seen that, with the real multivector calculus, it is possible to rewrite the Dirac equation. This fact has an important physical meaning; it is well known that for a spinor a rotation of it is necessary 4π in order that a spinor, which describes a quantum state of a fermion, comes back on itself, whereas a rotation of 2π will change the sign.

Now, in geometric algebra, a spinor is really a rotation, and its expression (see Equation 10.85), shows clearly what we have said. In fact, it transforms itself by composition: if one performs the rotation R', one finds $R \Longrightarrow R'R$, and for a rotation of 2π, $R' = -1$, and then $R \Longrightarrow -R$. Therefore, the famous change of the sign is due to the fact that R' acts only on one side of R, whereas for the vectors it acts on both sides (i.e., $a \Longrightarrow R'a\tilde{R}'$, and then, $a \Longrightarrow a$ for $R' = -1$). This behavior of fermions has been verified experimentally through neutron interferometry [11, 12]. This is important because experiments such as this throw light also on the role of gravitation in quantum mechanics [13]. The way in which a phase difference is induced between the two particle rays depends, in the first case, by it being induced from a magnetic field [11], and, in the second case, from the earthly gravitational field [13]; the result is similar in both cases, that is, one observes the usual peaks of interference pattern. In order to see how a magnetic field can change the phase of a spinor, one has to look at the precession of the spin, which, moreover, allows a beautiful application of geometric algebra.

Further, in Reference [14], it is shown that, in the context of the Dirac theory rewritten with the multivector algebra, the phase of the wave function shows an intrinsic geometrical meaning that leads to some considerations regarding the zitterbewegung. The free particle Dirac equation admits the plane waves solution:

$$\psi = R_0 \exp[-\gamma_2\gamma_1 p \cdot x/\hbar], \tag{10.80}$$

where R_0 is a constant Lorentz rotation and p the four momentum of the particle. Hence, the dependence on the coordinates is exclusively in the phase factor that in geometric algebra is a rotation spinor in the space-plane $\gamma_1 \wedge \gamma_2$. Consider now the system of the currents: because γ_0 and γ_3 are perpendicular to the plane $\gamma_2\gamma_1$, the velocity $e_0 = v = p/mc$ and the spin direction e_3 are constants (i.e., determined by R_0). This is not true for the vectors $e_1(x)$ and $e_2(x)$ that, on the other hand, have no analogy in the usual Dirac theory.
Because

$$e_2 e_1 = R_0 \gamma_2 \gamma_1 \tilde{R}_0 \tag{10.81}$$

we can also write

$$\psi = \exp[-e_2 e_1 p \cdot x/\hbar] R_0 \tag{10.82}$$

In that way we have shown a local rotation in the constant plane of the spin $e_2 e_1$ that one can also call the "phase plane." In conclusion, one has that the vectors $e_1(0) = R_0 \gamma_1 \tilde{R}_0$ and $e_2(0)$ rotate in the spin plane in a way that it is possible to define the phase of the wave

$$e_1(\tau) = \exp[-e_2 e_1 \omega\tau/2] e_1(0) \exp[e_2 e_1 \omega\tau/2] = \exp[-e_2 e_1 \omega\tau] e_1(0), \tag{10.83}$$

(notice that $e_1(0)$ anticommutes with the exp[ ])
where τ is the proper time of the electron ($p \cdot x = mc^2\tau$) and

$$\omega = 2mc^2/\hbar = 1.6 \times 10^{21} Hz \tag{10.84}$$

is the rotation frequency. This geometric fact can be interpreted as a manifestation of the zitterbewegung, the word with which Schrödinger [15] indicates the fluctuations (of the order of the Compton length) of the position of the electron. In that ambit the Compton length comes out as the typical length associated with the electron in a natural relativistic way; in fact, ω, as a rotation frequency, defines the limiting distance r (i.e., a real body cannot realize a rigid rotating system with frequency Ω for distance greater than c/Ω in the rotation plane) or the Compton radius

$$r_C = c/\omega = \hbar/2mc = 1.9 \times 10^{-11} cm \tag{10.85}$$

that we interpret as the ratio between spin and mass, the two fundamental dynamical quantities.

We have seen [1] that torsion is responsible for the fluctuation of the position $\Delta x^\mu \cong Q^\mu dS$, whereas the defect angle is due to curvature $\Delta\vartheta \cong$ RdS [16]. Because the curvature is linked to the mass and the torsion to the spin, we find, taking the time component

$$\begin{aligned} \Delta\vartheta/\Delta x^0 &\simeq RdS/Q^0 dS = (\chi \quad p \quad mc^2)/(c\chi \quad p \quad \hbar/2) \\ &= 2mc^2/c\hbar = \omega/c = 1/r, \end{aligned} \tag{10.86}$$

that the frequency ω, can be interpreted as the ratio between these two incertitudes. In conclusion, the quantum length r_C can be considered as the ratio between torsion and curvature, and this fact suggests, again, that Q and R can be considered as conjugate variables.

Appendix A: Commutator and Anticommutator

Now we must modify the commutation relations found in Section 10.3 (see equation 10.34) because, in that equation we have considered torsion and curvature trivectors, whereas in Quadratic Hamiltonian (see Section 10.4) we have to do with a bivector Q^α(torsion) and a trivector R^α(curvature), which are our canonical conjugate variables.

In other words, we try to improve the theory and we find important developments regarding the possibility to have commutation relations between torsion and curvature with some consequences regarding considerations about supersymmetry.

As we have seen, the commutators in Equation 10.34 allow a simple interpretation in agreement with the geometric content of the Dirac equation, but they are not entirely satisfactory because, in the Lagrangian (10.50) (and then in the Hamiltonian 10.52) the conjugate variables are the torsion bivector Q^α and the curvature trivector R^α, whereas in Equation 10.34 the commutation is made between trivectors.

In fact, working with geometric algebra, it is possible to have commutation relations between multivectors of different grades and not only between geometric objects of the same type. Yet, Equation 10.34 has meaning, and it is not to be rejected because we have started with a totally antisymmetric torsion, and then, the torsion trivector Q is the complete and single object of geometric algebra that contains the torsion field.

Now, the geometric product between a grade-r multivector A_r and a grade-s multivector B_s can be decomposed as (see Reference [10], page 10, Equation 1.36)

$$A_r B_s = (A_r B_s)_{r+s} + (A_r B_s)_{r+s-2} \ldots + (AB)_{|r-s|}. \tag{A.1}$$

If $r = 2$ and $s = 3$, being in four-dimensional space-time, we are left with two terms: $(AB)_1$ and $(AB)_3$. In other words, the geometric product BT, where B is a generic bivector and T a generic trivector, can be easily calculated remembering that every trivector T is the dual of some vector A (see Reference [13], page 107, Equations B1 and B2), i.e.,

$$T = iA, \tag{A.2}$$

where $i = \gamma_0\gamma_1\gamma_2\gamma_3$ is the unit pseudoscalar, and then, A is the vector dual of T (remember that the dual application, the multiplication by i, transforms an

r-vector in (4-r) vector). Moreover,

$$BA = B \cdot A + B \wedge A, \tag{A.3}$$

where the inner product

$$B \cdot A = (1/2)/(BA - AB) \equiv [B, A] \tag{A.4}$$

has grade 1, and the outer product

$$B \wedge A = (1/2)/(BA + AB) \equiv \{B, A\} \tag{A.5}$$

has grade 3.
Then, using also $iB = Bi$, one finds that

$$BT = BiA = iBA = i[B, A] + i\{B, A\} = [B, T] + \{B, T\}, \tag{A.6}$$

where the commutator between B and T has grade 3 (trivector), and the anticommutator has grade 1 (vector).

Finally, we can write:

$$Q^\alpha R_\alpha = [Q^\alpha, R_\alpha] + \{Q^\alpha, R_\alpha\}, \tag{A.7}$$

where

$$[Q^\alpha, R_\alpha] = (1/2) Q^{\alpha\mu\nu} R_{\alpha\mu}{}^{\sigma\rho} \gamma_\nu \wedge \gamma_\sigma \wedge \gamma_\rho \tag{A.8}$$

and

$$\{Q^\alpha, R_\alpha\} = (1/2) Q^{\alpha\mu\nu} R^{\rho}_{\alpha\mu\nu} \gamma_\rho. \tag{A.9}$$

As in Equation 10.15, in agreement with the uncertainty relation $\Delta Q \Delta R \geq L_{Pl}^{-3}$, we can put

$$[Q^\alpha, R_\alpha] = L_{Pl}^{-3} iu, \tag{A.10}$$

$$\{Q^\alpha, R_\alpha\} = L_{Pl}^{-3} v, \tag{A.11}$$

where u and v are unit vectors.

Therefore, given the conjugate variables Q^α and R^α, we have both commutator and anticommutator; we believe that this fact can be related to supersymmetry in the sense that one can treat simultaneously fermionic fields and bosonic fields if one considers, as in the second quantization procedure, the development of the fields in terms of creation and annihilation operators, which present analogies with the relations (A.10 and A.11) between torsion and curvature. However, this is an argument for future works.

References

1. H. E. Cartan, *Compt. Rend.* 174, 437, 593 (1922); *Ann. Sci. Ecole Normale* 41, 1 (1924).
2. A. Trautman, "Theory of Gravitation," preprint IFT/72/25, Warsaw University, read at the Symposium "On the Development of the Physicist's Conception of Nature," Miramare, Trieste, Italy, 1972.
3. V. de Sabbata, *Nuovo Cimento* 107*A*, 363 (1994).
4. B. K. Datta, V. de Sabbata, and L. Ronchetti, *Nuovo Cimento* 113*B*, 711 (1998).
5. V. de Sabbata and L. Ronchetti, *Found. Phys.* 29, 1099 (1999).
6. H. -H. v. Borzeszkowski and H. -J. Treder, *Ann. Phys.* (Leipzig) 46, 315 (1989).
7. H. -H. v. Borzeszkowski and H. -J. Treder, "Classical Gravity and Quantum Matter Field," in *Quantum Gravity*, ed. P. G. Bergmann, V. de Sabbata, and H. -J. Treder (World Scientific, Singapore) pp. 32–42 (1996); see also "Weyl–Cartan Space Problem in Purely Affine Theory," in *Spin in Gravity*, ed. P. G. Bergmann, V. de Sabbata, G. T. Gillies, and P. I. Pronin (World Scientific, Singapore) pp. 9–32 (1998).
8. H. -J. Treder, *Astr. Nachr.* 315, 1 (1994).
9. Yu Xin, *Astrophys. Space Sci.* 154, 321 (1989). see also "General Relativity on Spinorial Space-time" preprint.
10. D. Hestenes, *Clifford Algebra to Geometric Calculus* (D. Reidel, Dordrecht, Holland, 1982).
11. V. de Sabbata and M. Gasperini, *Introduction to Gravitation* (World Scientific, Singapore 1985).
12. B. K. Datta, R. Datta, and V. de Sabbata, *Found. Phys. Lett.* 11, 83 (1998).
13. B. K. Datta and V. de Sabbata, "Hestenes' geometric algebra and real spinor field," in *Spin in Gravity*, ed. P. G. Bergmann, V. de Sabbata, G. T. Gillies, and P. I. Pronin, (World Scientific, Singapore, 1998), pp. 33–50, Eq.(6.1).
14. B. K. Datta, "Physical theories in space-time algebra," in *Quantum Gravity*, ed. P. G. Bergmann, V. de Sabbata, and H. -J. Treder (World Scientific, Singapore, 1996), p. 54–79.
15. A. Lasenby, C. Doran, and S. Gull, *Found. Phys.* 23, 1295 (1993).
16. R. T. Rauch, *GRG* 14, 331 (1982).
17. R. T. Rauch, *Phys. Rev. D*25, 577 (1982).
18. R. T. Rauch, *Phys. Rev. D*26, 931 (1982).
19. J. A. Schouten, *Ricci-Calculus* (Springer-Verlag, Berlin, 1954) p.141.
20. H. Euler and B. Kockel, *Naturwissenschaften* 23, 246 (1935); H. Euler, *Ann. Phys.* (Leipzig) 26, 298 (1936); W. Heisenberg and H. Euler, *Z. Phys.* 98, 740 (1936).
21. H. -H. v. Borzeszkowski, B. K. Datta, V. de Sabbata, L. Ronchetti, and H. -J. Treder, *Found. Phys.* 32, 1701 (2002).

Index

C

D

H

N

O

T

U

V

W

X

Z